A Guide to Writing in the Sciences

Andrea A. Gilpin and Patricia Patchet-Golub

A Guide to Writing in the Sciences is an introduction to the basic principles and conventions of science writing. A practical handbook to which students can refer at various stages of their studies, this manual evolved from the authors' own experiences while teaching writing to undergraduate science students. Andrea Gilpin and Patricia Patchet-Golubev discovered that, while there are many manuals for scientific writing, most are aimed at professionals or graduate specialists and are of limited use to the majority of students. This introductory guide fills an important gap in the literature.

The book provides clear and concise explanations of the basic elements of scientific writing required of students. The various scientific genres are detailed, with an emphasis on lab reports and research essays. However, scientific articles, poster presentations, proposals, and even essay exams are also covered. Similarities and differences among these genres are underlined in order to illustrate structural principles and to provide practical writing tips. A special chapter outlines the elements of grammar and punctuation that are fundamental to all good writing; it also details some key points about scientific style in particular.

The authors organize their material with helpful lists, examples, and writing and research tips. There are several appendixes (including a classification chart of organisms and an international units conversion chart), an index, and a selected bibliography on science and technical writing. Authoritative and accessible, this guide is an invaluable addition to any science student's library.

ANDREA A. GILPIN received her PhD in the Department of Botany, University of Toronto, in genetics and molecular biology.

PATRICIA PATCHET-GOLUBEV has been a writing instructor for eight years at the University of Toronto and has taught in the Continuing Education Department at Ryerson Polytechnic University.

A Guide to
Writing in the Sciences

Andrea A. Gilpin

Patricia Patchet-Golubev

UNIVERSITY OF TORONTO PRESS
Toronto Buffalo London

© Andrea A. Gilpin and Patricia Patchet-Golubev
Published by the University of Toronto Press 2000
Toronto Buffalo London
Printed in Canada

ISBN 0-8020-8366-8

Printed on acid-free paper

Canadian Cataloguing in Publication Data

Gilpin, Andrea A., 1969–
  A guide to writing in the sciences

  Includes index.
  ISBN 0-8020-8366-8

  1. Technical writing.   I. Patchet-Golubev, Patricia.   II. Title.

  T11.G54 2000   808′.0665   C00-931240-4

The University of Toronto Press acknowledges the financial assistance to its publishing program of the Canada Council for the Arts and Ontario Arts Council.

University of Toronto Press acknowledges the financial support for its publishing activities of the Government of Canada through the Book Publishing Industry Development Program (BPIDP).

Contents

Acknowledgments

We would like to acknowledge the following individuals, who generously read and expertly commented on different versions of the manuscript: Nancy Dengler, Carolyn Hutcheon, Steve Mezyk, Nicholas Schisler, and Wayne Snedden. Our gratitude, as well, to Anne Cordon and the BIO250 team, and the JLM349 team, who let us "test run" a preliminary draft on their undergraduate students and provided us with invaluable suggestions along the way. Also, Margaret Procter deserves particular mention for her continued support and encouragement on all levels. In addition, a special thanks to Gregg Nyhus for his advice and help from beginning to end and, last but not least, to Felix, Madeleine, and Rebekah Golubev – for their needed good humour and patience.

A Guide to Writing in the Sciences

Introduction

The idea for this guide grew out of our collaborative teaching experiences in an innovative project at the University of Toronto. Beginning in January of 1999, a non-credit course, "Writing for Scientists," teamed writing specialists up with scientists to teach undergraduate students in their third and fourth years. These students, from a variety of backgrounds, in terms of both discipline and linguistic ability, had one thing in common – they were strongly motivated to improve their writing skills, especially in the various genres of scientific writing.

As co-instructors for one section of this course, we enjoyed the challenge of interdisciplinary teaching, and learned a great deal in the process. Drawing on a variety of resource materials to supplement our course (in both paper and electronic form), we quickly became aware that a practical science writing guide, geared specifically to the level and demands of undergraduate students, was distinctly lacking. Consequently, the following summer, with the generous assistance and support of many individuals, we set out to produce such a manual. The result is the guide that you are now reading.

We initially aimed our guide at an undergraduate science readership. In fact, the first year that we printed it, we were fortunate in having a large second-year biology class "test run" and evaluate it for us. Thus, the writing genres on which we focus (primarily, the lab report in chapter 1; second, the research essay in chapter 3; and, finally, the proposal and poster presentation in chapter 2) are the ones most likely to be encountered by undergraduate students from the first year of their studies onwards.

In addition, many students and other readers advised us that such a guide would do well to include a chapter dealing with the particular writing demands posed by essay exams. Increasingly, in many post-secondary settings, science students are being expected to develop their analytical (critical reading, thinking, and writing) skills. As a result, instructors are often turning to essay-type exams (as an alternative to multiple-choice ones) to evaluate their students. Hence, we developed chapter 4, which outlines study strategies and skills for both short and long essay exams. Last, but certainly not least, the grammar and style tips covered in chapter 5 will, we believe, be indispensable for all students, both those for whom English is a second (or third) language and native speakers.

Numerous professors and graduate students (who have generously given their time to read through and comment on our manuscript) have also pointed out that our guide may indeed be very useful for junior (and even senior) graduate students. So, with these comments in mind, we worked to expand and enlarge upon some aspects of the guide to address the needs of an even broader audience. Essentially, we have directed our guide to any and all who wish to improve their writing skills, both basic and advanced ones.

WRITING: SOME INTRODUCTORY REMARKS

Any kind of writing – be it a poem, newspaper editorial, business memo, novel, history textbook, or scientific article – is, first and foremost, a form of *communication* between a writer and his or her readers. Writing represents a *means* of conveying information, ideas, facts, emotions, theories, arguments – and hundreds of other things – to somebody else, a reader or group of readers. So, as a writer, you should always retain a clear and focused idea of exactly who your intended readers are.

Another key point about the process of writing is that it is intrinsically connected to the process of thinking; thus, clear and effective thinking usually brings about clear and effective writing (and vice versa). This point cannot, we think, be overstated. Too many people wrongly believe that both learning and thinking are divorced from writing (first we learn something, and then we write it down); in fact, writing is a key part of the learning process.

Finally, as beginning writers (and scientists), you need to realize that writing skills are not mysterious gifts that some people possess and others do not. With steady practice and application, almost anybody can learn to perfect most writing skills. The more you write, the more you will master these skills, and the less anxiety you should experience in the process. If our guide helps somewhat to alleviate this anxiety, we will have accomplished at least one of our goals.

Chapter One

The Laboratory Report

LABORATORY REPORT, PRIMARY RESEARCH ARTICLE, REVIEW ARTICLE – WHAT ARE THE DIFFERENCES?

The organizational structure of a **lab report** is rigidly defined, unlike that of a regular academic essay, scientific or otherwise. The lab report is divided into seven main sections: Title, Introduction, Materials and Methods, Results, Discussion, Abstract, and References. Some refer to this as the IMRAD (Introduction, Methods & Materials, Results, and Discussion) formula for scientific writing.

A lab report is structured similarly to a **primary research article** (according to the IMRAD formula). Nevertheless, both the scope and complexity of a primary research article's Introduction, Results, and/or Discussion sections are generally greater than those of a lab report. For example, the Introduction of a research article, unlike that of a lab report, as well as providing background for the problem and a rationale for studying it, might also summarize the experiments previously attempted in the form of a literature review and outline the specific experimental approach adopted. In addition, the audiences for the two genres differ – research scientists inform the larger scientific community, primarily their specialized peers, of their findings in primary research articles, while lab reports (generally unpublished) are written for instructors or student colleagues. Further, primary research articles can be found in specialized journals such as *Cell, Journal of Biological Chemistry, Journal of Physics*, and *Science*.

A primary research article itself differs significantly from a **review article**. The latter summarizes or synthesizes and, more important, evaluates the concepts and / or results from several research articles. Thus, authors of review articles compare, contrast, and interpret the work of others on a particular question or topic. Not based on primary research, these articles thus usually contain no graphs and tables in their Results section. Any figures that do appear in a review article generally pull together information from various sources rather than present primary data. Because of their different purpose, review articles do not typically follow the IMRAD formula of lab reports and primary research articles. Some examples of journals that print review articles are *Trends in Biochemistry, Current Opinion in Chemical Biology, American Scientific*, and *Annual Review of Ecology and Systematics*.

In this chapter, we will focus on the lab report by discussing each section of the report separately and giving you some useful tips on how to get started. Because of the formulaic structure of the lab report (IMRAD), you might conclude that organization is never a problem. This, however, is not necessarily the case, since each section of the paper still requires a coherent and logical internal structure. Clear presentation of all the data supporting or refuting your hypothesis is essential in order to persuade your readers.

In this chapter, we order the sections as they actually appear in a lab report; such a presentation does not imply, however, that you should necessarily *write* the sections in this order. Professional scientists often write research articles in the following order (try it for yourself and see if it works for you). They begin by writing the Results section, the real center of the report, which contains the most important information. The actual findings presented in this section and how they are organized will often determine what is included within the Discussion and Introduction. Thus, after finishing a draft of the Results section, writers generally have a good idea of how to go about interpreting their data in the Discussion section. Then, only after completing the Discussion, is it usual to compose the Introduction, since now they know exactly what it is that they are providing a context or background for. Next, the Materials and Methods can almost write itself because the Results can serve as a kind of checklist. The final section written is the Abstract; since the Abstract is a summary of the whole report, it can be written only after the whole report is finished.

In the past, references were tedious to type out because of all the intricate formatting. Now, however, because of word processors and reference managers, the task is much easier. We strongly recommend that you use a reference manager if at all possible. Otherwise, try to list your references as you write your report; that way, the references will be complete as soon as the report is. Finally, it makes sense to write the title at the very end, since by then you should have the "whole picture" in mind.

TITLE

Titles indicate the purpose of the investigation and may also describe the methods or results. Be as brief as possible, while also remaining specific. A good approach to writing an effective title is to list the key ideas or concepts within your report. You can then make up several different titles from various combinations of these words. Although you may decide *not* to include your key result in the title, one advantage of doing so (especially if it is an unusual result!) is to "hook" or attract readers. Sometimes, omitting the result from the title may also pique potential readers' curiosity. Both approaches are acceptable; your choice will depend upon personal preference and the actual key result of your paper. With a little work, a suitable title will soon come together.

> *For example:* Some DNA was isolated from Mr Jones's epidermal cells and, by using PCR, we were able to prove that Mr Jones is, indeed, Billy's father.
> *List of key words:* DNA isolation, PCR, Mr Jones, paternity, Billy, epidermal cells
>
> *Possible title:* A PCR approach using epidermal cells determined the paternity case of Mr Jones [result not stated].
> *Alternative:* A PCR approach using epidermal cells demonstrated that Mr Jones is Billy's father [result stated].

INTRODUCTION

Your Introduction should "frame," and provide background for, the specific question or hypothesis of your investigation. It generally consists of four key elements. First, you ought to introduce the general area of science your report deals with and justify the importance of research in this area. Second, you should review key concepts integral to understanding the paper. Third, you could develop the scientific context – in other words, explain what is already known or unknown about this question, as reflected in the scientific literature. Fourth, you must state the hypothesis or purpose of your investigation.

But what, exactly, is a hypothesis? What, in regular academic essays, is called a "thesis" or "thesis statement" is referred to as a "hypothesis" in scientific writing. The Merriam-Webster Dictionary defines "hypothesis" as "a tentative assumption made in order to draw out and test its logical or empirical consequences." In other words, **a hypothesis is based on a prediction about a certain process or phenomenon**, which scientists attempt to test. The hypothesis often determines which experiments will be performed to test it. For example:

> *Hypothesis:* "The known genes responsible for cysteine biosynthesis in cyanobacteria are transcribed only during sulfur deprivation. Since cysteine is a critical amino acid, my hypothesis is that alternative methods of producing cysteine must exist ..."

In this example, the author clearly states the observation followed by his/her hypothesis.

Some exploratory approaches commonly used in scientific writing have a different type of statement of intent than the formal (and traditional) hypothesis outlined above. This kind of statement, called a **focus statement**, outlines an investigative approach to a problem on the basis of some observation, but the focus statement itself is not tested as is a hypothesis. Instead, the focus statement describes a series of steps to explore a certain area that can be neither "accepted" nor "rejected." Rather, the experimental approach is more one of "look-see."

1. *Focus Statement:* "The genetic approach of a mutant screen was performed to isolate mutant plants defective in their responses to carbon dioxide."

In this focus statement, the author (a botanist) is interested in elucidating the response mechanisms to carbon dioxide in plants. As such, he/she attempts to isolate components of the carbon dioxide pathway and to analyze the mutants that come out of the screen. The focus statement will not be formally tested or rejected.

2. *Focus Statement:* "Efficient reaction conditions for the production/synthesis of deoxyglucose synthesis were determined."

In this focus statement, the author (a synthetic chemist) articulates his/her attempt to discover the optimal conditions to synthesize deoxyglucose (by altering temperature, pressure, pH, etc.) to a complete reaction (such that yields near 100% are achieved). The approach is focused despite the lack of a traditional, testable hypothesis.

GETTING STARTED ON THE *INTRODUCTION*

Write out the overall, skeleton structure of the introduction in point form:

Paragraph 1 – Introduce the topic and indicate what is generally known about it.

Paragraph 2 – Write down the concepts that need to be reviewed to understand the question described in the paper.

Paragraph 3 – Review the current literature on the topic in order to provide context for the hypothesis. You might also point out what is *not* known (such an approach may lead effectively to your hypothesis).

Paragraph 4 – Determine whether or not your study is exploratory and state either your hypothesis or your focus statement (preferably in the last paragraph of the introduction). Indicate the importance of your work and how it will contribute to current scientific knowledge. Here, you could also briefly justify your methodology.

MATERIALS AND METHODS

This section presents, in a logical order, the materials used in the experiment and the procedure(s) by which the experiment was performed. You should describe or explain the materials and methods used in your experiment in such a way that your reader could repeat the experiment. Writing the Materials and Methods is not always a simple task, for, contrary to popular belief, it is not necessarily easier to describe something than it is to analyze or evaluate it. Good descriptive writing requires time, thought, and thorough attention to detail.

When writing descriptions for your Materials and Methods, consider carefully both the audience and the purpose. Ask yourself the following question: What does my audience already know? (For example: if you are describing a biochemical assay, decide whether or not your audience knows what an assay is.) It may occasionally be sufficient to refer your reader to the lab manual for the entire Materials and Methods, but for many courses, you may have to write your own Materials and Methods.

With respect to materials, here are some tips. For machine names, include the company name and model number, as well as the settings used. If you ordered plant/animal/bacterial matter from a stock center, include the company name and the strain number you used.

> Exogenous DNA was introduced into *E. coli* by electroporation (Gibco BRL Electroporator, at settings of 1 ms, 2.4 kV pulse).

> *Arabidopsis thaliana* ecotype Columbia seeds were obtained from Smith Seeds and grown on minimal plant medium.

With respect to methods, you should *not* typically enumerate procedures as a set of step-by-step instructions but, instead, describe the process in sentence and paragraph form. Use the third person, passive voice (e.g., "The DNA sequence was determined") in order to emphasize the procedure and *not* the person doing the experiment. Some instructors prefer that you set up your methods as a flow chart. Check with your professor or instructor to determine which format is most appropriate for your report.

State any key assumptions you made in working out the experimental design (lengthy or complex assumptions, however, may be more appropriately dealt with in the Discussion). Describe what you did and explain how, providing sufficient details for readers to assess the reliability of your methods. Cite references to methods published in accessible scholarly journals instead of repeating the details of these methods. If applicable, provide any alterations you have made to the protocol.

> "The hydrostatic equilibrium was determined as previously described by Van Thielen *et al.* (2000)."

> "RNA was isolated as previously described by Nyhus *et al.* (2000) with the following modifications: pre-hybridization solution contained salmon sperm DNA (10 mg/ml) as a blocking agent and the hybridizations were performed at 65°C."

Indicate the methods of statistical analysis used (if applicable) and clarify which one was employed in each part of your investigation. If you used many statistical methods, repeat each one in the appropriate place in Results. When a particular statistical technique has more than one form, state which form you used, including any arbitrary levels such as probability (e.g., $p > 0.05$). Explain any complex statistical methods and why you chose them. If you used a particular computer program to analyze your findings, indicate which one and which version; also identify your specific methods of analysis. Cite these statistical methods in the References section. For voluminous statistical tables and/or raw data, an appendix at the end of your report may be more appropriate. Finally, ensure that you describe precisely and concisely what you did. Try not to overlook any details. If possible, have someone who is familiar with your investigation proofread your Materials and Methods to ensure that you have not omitted any important details.

KEY TIPS FOR WRITING *MATERIALS AND METHODS*

- Use the past tense and passive voice. For example: "The sample was analyzed."

- Ensure that all measurements are in acceptable SI units. For example, the correct SI unit to measure light is µmol $m^{-2}s^{-1}$, not µEinstein. To measure the amount of work done, Joules are the proper SI units, not Calories (see appendix 3).

- Give the chemical IUPAC names for chemical substances. Do not use trade or common names unless they are the most accurate method of identification. For example, instead of "bleach," write "sodium hypochlorite."

- Assign concentrations in percentages (acceptable in biology) or moles/volume (typically used in chemistry). For example, it is meaningless to indicate that you used 10 µl of an antibiotic unless the reader knows the concentration of the stock solution. Thus, "an ampicillin concentration of 50 µg/ml of solution" is more descriptive.

- For species names, assign the full Latin name (in italics), including the source, strain, breed, cultivar or line, as appropriate. Include the authority responsible for the nomenclature you use (if applicable). When you first use a species name, spell it out in full with the genus name capitalized and the species name in lower case. Subsequent uses can be shortened (e.g., *Escherichia coli* can be shortened to *E. coli*, *Pseudomonas syringae* to *P. syringae*; see appendix 2).

- In general, mutant gene abbreviations have three letters written in lowercase italics (*fer*); the wild type allele abbreviation should be written in uppercase italics (*FER*). The same nomenclature applies to spelling out the entire gene name. Protein products have the same nomenclature without the italics. Note that this gene convention is not used for all systems, so verify exactly how the system you are using is denoted.

- Phenotypes may be denoted as the gene abbreviation without italics, with the first letter capitalized. A superscript "+" refers to the wild type phenotype, while a superscript "–" refers to the mutant phenotype (Fer^+ – wild type allele; Fer^- – mutant allele).

- When first mentioning enzymes, provide their systematic name and Enzyme Commission numbers (EC) (e.g., ACC synthase [EC 4.4.1.14]); use the trivial names subsequently.

- Give the recognized name of restriction enzymes and the source from which you obtained them. Italicize the restriction enzyme name but not the strain number (e.g., *Eco*RI, *Bam*HI).
- Include the name of the supplier with chemicals or machines that are not standard.
- Include the sex, age, and growing conditions of all organisms, as well as their genetic, physiological, and dietary status, and a description of the conditions under which they were kept.
- For geographical areas, use accepted names and spellings, following recognized authorities for geological nomenclature.
- Include temperatures, the pH of buffers, the speed of centrifugation (in x g, *not* rpms!), and the duration of any treatment.

RESULTS

Professional scientists generally read the Abstract of a paper before the paper itself. They may then glance at the Introduction and Materials and Methods, but they usually concentrate on the Results, since this section contains the "heart" of the article. All other sections are subordinate to this one; here, readers obtain the most significant and useful information. Scientists may question the approach in the Methods or dispute the interpretation of the results found in the Discussion, but the Results themselves should give clear and unambiguous data on the basis of which the reader may construct his/her own evaluation. Therefore, you need an effective strategy for presenting your results.

The Results section should accurately report and describe (in the past tense) all the data you have collected, including sufficient detail to justify the conclusions in your Discussion section. **Do not interpret the data here**, but instead provide statistical summaries of results (present voluminous numerical data as percentages, averages, totals, etc.). Significant observations and trends should be highlighted; make it clear how they relate to your argument, without further interpretation.

"The lipoxygenase enzyme activity appeared to decrease over the five-day period." (Note: The author emphasizes, but does not interpret, the trend.)

If certain results do not agree with your hypothesis, mention this contradiction, but do not elaborate on it in the Results (e.g., "These results are inconsistent with the experimental hypothesis.")

Although the Results and Discussion often appear as separate sections, some science editors maintain that the two should appear together. Whether you present the results and discussion separately or together depends on your particular experiment and the purpose of your report. Your instructor may give you some indication about what he/she wants. In general, if you do not intend to discuss the findings in detail or believe that the reader needs the interpretation of an earlier result to understand a later one, then it might make sense to combine the two sections. However, if you think it useful to discuss the results as a whole after they have been reported, present the two sections separately.

Tables, Figures, and Legends

Tables, figures, and figure legends are key elements of every scientific paper, allowing the reader to absorb data quickly in a condensed and logical form. Readers often evaluate the rationale of your approach and the soundness of your interpretation on the sole basis of your figures and tables. So make sure that they work to your advantage. Tables and figures should be able to stand on their own, without reference to the text; nevertheless, you must also refer within the text of the Results section to each and every table and figure presented. Another important point: descriptive, clear, and concise figure legends are essential to guide readers through the data in your figures.

A table consists of a title, column and row headings, a field (the rows and columns containing the data), and, usually, explanatory notes. A table's title should be informative, emphasizing what the table shows, and be located above the table. An example of a good title for a table is as follows:

"Increasing incidence of breast cancer in three independent areas in Texas, 1995–2000."

This is preferable to "Breast cancer in Texas."

A table must contain enough data to justify its existence, but its structure should be as straightforward as possible. Ask yourself which key message you want readers to extract from your tables, and ensure that they clearly convey this message.

Sample tables

Table 1: Comparison between shoots and roots.

	Length of Shoots 1% Sucrose	Length of Shoots 3% Sucrose	Length of Roots 1% Sucrose	Length of roots 3% Sucrose
Wild type	10.5	17.5	16.0	36.2
Mutant	15.95	26.7	20.0	40.3

Table 1 is poorly constructed. The data have not been clearly presented; therefore, it is not immediately clear what the author expects the reader to extract from the table. First, the title is not descriptive. Second, precise units of measurement are missing. Third, the significant digits are not consistent. Finally, a range of error has not been indicated. Compare Table 1 to Table 2. In the latter, the title is comprehensive and clear, the reader is given the units of measurement in mm, the range of error is supplied, and the headings are more descriptive. Thus, Table 2 presents the data much more successfully than does Table 1.

Table 2: Comparison of the shoot and root length of tomato plants when grown on media containing various sucrose concentrations.

	Shoots (mm)		Roots (mm)	
	1% Sucrose	3% Sucrose	1% Sucrose	3% Sucrose
Wild type	10.50 ± 0.5	17.50 ± 0.8	16.09 ± 0.6	36.21 ± 0.9
Mutant	15.95 ± 0.3	26.71 ± 1.2	20.01 ± 0.8	40.34 ± 1.1

Alternatively, you may choose to represent your data in a graph or a figure. Graphs exist in several varieties, including line graphs (showing dynamic relationships between two or three variables over time), bar graphs (useful in the case of one or two variables, or when findings can be subdivided and compared in different ways), histograms (indicating the frequency distribution for each variable), and pie charts. Figure legends belong beneath the figure. Other kinds of figures used in a Results section include maps, algorithms, flow charts, models, and photographs.

Sample figures (bar graphs)

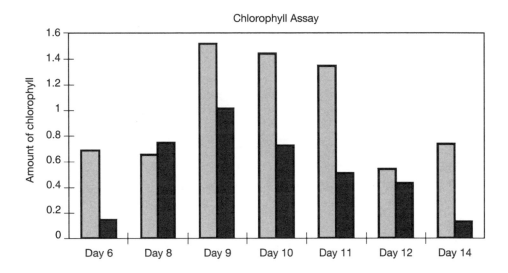

Figure 1: Amount of chlorophyll in leaves.

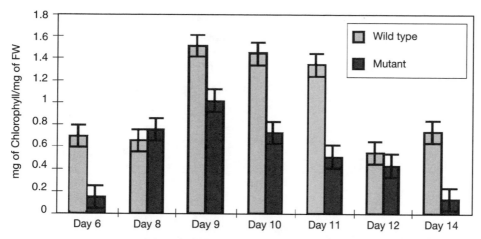

Figure 2: A comparison between the wild type and mutant plants in the amount of chlorophyll in tomato leaves at various developmental stages.

The figure legend of Figure 1 describes the data poorly. The graph indicates no units, contains no error bars, and uses weak colors (colors that appear to contrast sharply on a monitor may not look so contrasting on a black and white printer!). Compare Figure 1 with Figure 2, which represents an improved figure in all respects. Figure 2 could easily be printed on a black and white printer, and the adjacent bars would be clearly noticed. In addition, the error bars in Figure 2 clearly reveal the significance of the data.

Some things to keep in mind when making figures:

- Make sure all axes are well labeled with the correct units. Where appropriate, ensure the units on the x and y axes are comparable.
- Create your bar lines so that they can be clearly seen.
- If you use colored lines, be certain that your lines are thick enough to be visible in every color (especially yellow!).
- If you use symbols for points on line graphs such as triangles, circles, and squares, be certain that they are large enough to be visible.

Table or Figure?

In general, it is preferable to represent your data in the form of a figure, since figures usually illustrate relationships between variables better than do tables. Tables are also often cumbersome and hard to read. Here are two points to consider when making your decision:

- Figures and graphs are ideal for describing trends.
- Tables can display numeric data that can be compared easily with other data.

GETTING STARTED WITH THE *RESULTS*

- Use the past tense.
- Examine each piece of relevant data and decide how to represent it most effectively. Try several approaches before settling on the best one.
- Analyze your data statistically. If possible, show significant differences and/or include in your graphs and tables either standard error or standard deviation in the data points.
- Focus your Results section by writing out your figure legends.
- Place your figures in a logical order, which can then become a template for the text of your Results and Discussion.
- Begin writing your results by describing each figure and its significant finding(s). Do not interpret or provide context in this section!
- Within the text, refer to the result(s) presented in a figure or table, not to the figure number. For example, write: "The amount of force increased proportionately to the angle of the slope (Figure 1)," instead of "Figure 1 shows that the force increased as the slope increased."
- Discuss each result together with its interpretation only if you are combining the Results and Discussion into a single section.

DISCUSSION

This section is the least structured of all. Here, you are expected to examine, discuss, analyze, and interpret your data, commenting on their significance. Analyze the data in the context of the experiment's purpose. It is a good idea to restate the experimental outcome, to indicate whether the outcome was expected or unexpected, and to discuss those elements that influenced the result(s). Try to make logical and clear deductions in your discussion, referring to the data and findings in the Results section.

In the Discussion, you answer the general question (in the present tense): "What do these findings mean?" More particularly, you are dealing with the following specific questions:

- What (if any) are the major assumptions implied in your experiments?
- Do the results meet the experiment's objectives?
- Do the results agree with expected results or with previous findings as reported in the literature? If not, how can you account for the discrepancy between your results and those expected? What may be wrong with published data that contradicts your results?
- What, if anything, may have gone wrong during your experiment, and why?
- Was there any source of error?
- Could the results have more than one explanation?
- Did the procedures you used help you to accomplish the purpose of the experiment?
- Does your experience in this experiment suggest a better method for the next time?
- Could another experiment be devised that may corroborate your result?

Explain what is unique about your experimental approach and indicate why your results are significant, without making extravagant claims for them. It is possible, as well, to theorize as to what the next steps of the experimental process might be. In addition, discuss other relevant results and hypotheses, always distinguishing between facts and speculation. When you present your experiment's conclusions (usually in the final paragraph of your Discussion), use forceful verbs such as "show" or "indicate." But when you are speculating about something, insert auxiliary verbs such as "might" or "could." Remember also that statistical analyses by themselves can never absolutely prove anything; they only indicate that the opposite is unlikely and give a measure of how unlikely it is (i.e., $p < 0.05$).

Concentrate on the main lines of your argument, avoiding repeated reference to every detail of your work. A common error is rewriting the Results into the Discussion.

GETTING STARTED WITH THE *DISCUSSION*

- Jot down, in point form, answers to the above questions. Outline only the key points relevant to your hypothesis or focus statement.
- One possible way to open the Discussion is to begin with your hypothesis or focus statement. Try to ensure that the question posed is still recognizably the same. Develop and interpret the answer to this question throughout your Discussion.
- Discuss your findings in a logical order (usually this order will be similar to the one used in the Results).
- Support, as much as possible, your interpretations with what is known from the scientific literature. If nothing (or little) is known, state this (you will thus emphasize the importance of your own research).
- After developing the argument provided in your Discussion, write a conclusion near its end.
- Conclusions are not meant to summarize the Discussion, but rather to serve as deductions that follow clearly from the evaluation of the data. They should represent a strong and final statement about your work.
- Do not make your conclusions too strong for the data and supporting argument.
- Following the conclusions, you may briefly indicate which questions remain to be answered and suggest possible approaches for answering them.
- Reference sources properly throughout your Discussion. Provide solid research and support for all the ideas you develop in your Discussion (see References section in this chapter).

ABSTRACT

An abstract is a brief, yet comprehensive, summary of your report, without added interpretation or criticism. It should convey the most significant information about your research, especially the results, and it belongs at the beginning of your report, before the Introduction. In primary research articles, readers often decide, on the basis of the Abstract, whether or not to read the full article. Since the Abstract summarizes the entire paper, you should write it last, after having completed all other sections. A single paragraph of approximately 250 words usually suffices. The Abstract should

1. be concise, but not highly abbreviated;
2. assume the reader has some knowledge in the subject area;
3. state the question or hypothesis;
4. present the approach used to answer this question;
5. report the most important results (but not cite figures or tables); and
6. indicate the main conclusion(s) near the end.

Since they are so condensed, abstracts can be difficult to write. The language must be terse and to the point. Some key phrases to link and introduce ideas are:

The main purpose of this study was to ...
To address this question we used ... approach (rationale)
The main findings suggest that ...
Therefore, these results support the conclusion that ...

GETTING STARTED ON THE *ABSTRACT*

- Announce the question under investigation with as little background information as possible, including only the main reason for the study.
- Name significant features of the model system (organism), experimental method, or approach.

- Report the principal findings (including statistical significance levels) without interpretation.
- Present your main conclusion.
- Keep sentences short and simple, but complete. Do not use sentence fragments. Make each sentence deal with only one topic and exclude irrelevant points.
- Ensure that sentences follow one another logically (the flow and emphasis of ideas/points often parallel those in the report).
- Use active, not passive, verbs and the past tense to report findings. (e.g., "SAG:kn1 tobacco plants showed a marked delay in leaf senescence but otherwise developed normally.")
- Never refer to information that is not presented in the paper.

Sample Abstracts

Inbreeding load, average dominance and the mutation rate for mildly deleterious alleles in *Mimulus guttatus*

J.H. Willis

The goal of this study is to provide information on the genetics of inbreeding depression in a primarily outcrossing population of *Mimulus guttatus*. Previous studies of this population indicate that there is tremendous inbreeding depression for nearly every fitness component and that almost all of this inbreeding depression is due to mildly deleterious alleles rather than recessive lethals or steriles. In this article I assayed the homozygous and heterozygous fitnesses of 184 highly inbred lines extracted from a natural population. Natural selection during the five generations of selfing involved in line formation essentially eliminated major deleterious alleles but was ineffective in purging alleles with minor fitness effects and did not appreciably diminish overall levels of inbreeding depression. Estimates of the average degree of dominance of these mildly deleterious alleles, obtained from the regression of heterozygous fitness on the sum of parental homozygous fitness, indicate that the detrimental alleles are partially recessive for most fitness traits, with (h) over bar similar to 0.15 for cumulative measures of fitness. The inbreeding load, B, for total fitness is similar to 1.0 in this experiment. These results are consistent with the hypothesis that spontaneous mildly deleterious mutations occur at a late > 0.1 mutation per genome per generation.

(Taken from: Willis, J.H. (1999) Genetics 153: 1885–1898. Reproduced with permission.)

In the above abstract, the author opens with the purpose of the study, thus presenting his implied hypothesis. The second sentence outlines what is known in the field; the author then explains the methodology used. Next, he briefly mentions his results and concludes with an explicit statement of his hypothesis.

Microarray analysis of Drosophila development during metamorphosis
Kevin P. White, Scott A. Rifkin, Patrick Hurban, David S. Hogness

Metamorphosis is an integrated set of developmental processes controlled by a transcriptional hierarchy that coordinates the action of hundreds of genes. *In order to identify and analyze the expression of these genes*, high-density DNA microarrays containing several thousand *Drosophila melanogaster* gene sequences were constructed. Many differentially expressed genes can be assigned to developmental pathways known to be active during metamorphosis, whereas others can be assigned to pathways not previously associated with metamorphosis. Additionally, many genes of unknown function were identified that may be involved in the control and execution of metamorphosis. The utility of this genome-based approach is demonstrated for studying a set of complex biological processes in a multicellular organism.

(Reprinted with permission from White, K.P. *et al.* (1999) Science 286: 2179–2184. Copyright 1999 American Association for the Advancement of Science.)

Here, the authors begin the abstract by describing the importance of the field under investigation. The italicized phrase indicates the focus statement of the research; the following clause ("high-density DNA microarrays

... were constructed") reveals the methodology that was used. Next, they explain how to use a new approach to identify new genes involved in metamorphosis. They conclude the abstract by stating how this new approach may be useful for many researchers.

REFERENCES

The Ethical Responsibilities: Plagiarism and Other Dilemmas

The citing of sources is crucial in scientific writing, as it is in all academic writing. But why do we really need to reference? First and foremost, we document our sources of information because of our ethical and legal responsibilities as authors (whether in the sciences or other disciplines) to acknowledge both the ideas and the research of others. You have probably heard something about the serious offence of "plagiarism," which refers to one's attempt to represent as one's own the ideas or expressions of the ideas of others. Thus, whenever one purchases essays (or, for that matter, any other written material) or copies whole passages, sentences, or even phrases (either verbatim or by paraphrase) from books, articles, or unpublished manuscripts without acknowledging the sources of this material, one is committing plagiarism. Likewise, falsifying any of your data or observations (to make them "fit" your hypothesis, for example) is unethical and should never be done. Every academic institution has its own way of dealing with plagiarism; in all cases, the penalties are severe.

Nonetheless, you document your sources not only to avoid the dire penalties of plagiarism; you do it also to show your reader that you have a solid and clear understanding of what is current and known in your field of study. Referencing also helps your readers by pointing them towards any material that you have found to be useful. So giving credit to others actually increases your own credibility as well!

Referencing can provide some problems for many beginning authors. Part of the difficulty is the actual referencing format (we'll get to that soon!). But, even knowing when and what to reference (in order to avoid the pitfalls of plagiarism) can be somewhat tricky. Junior (and even senior) under-graduates often wonder exactly how much to reference, since they think they know so little about the question or topic before they begin researching it that they should reference everything. The general rule is that you do not need to reference anything that is "common knowledge" in the field. "Common knowledge" applies to both facts and interpretations. When in doubt about what is considered "common knowledge" in your particular discipline, consult with your instructor or professor. As a rule of thumb, then, you should document the following:

- Any direct quotations, paraphrases, and summaries. Do not waste space on long quotations. In scientific writing, direct quotes are rare.

    In conservation biology, it is necessary to "maintain the distinctiveness of unique evolutionary lineages and genetic diversity within these lineages" (Ashley, 2001).

- Specific facts used as evidence for your interpretation (if these facts are not considered "common knowledge"). Any newly published or little-known findings, conclusions, or theories should thus be cited.

    Other recent researchers (Lorch, 1999; Snedden, 2000; Karn, 1998) have confirmed that PCR provides rapid and accurate analysis from many different tissue types.

- Distinctive or authoritative ideas, whether you think they are correct or not.

    Pearlman (2001) maintains that a mere eighteen months from now, after the first generation of modified hemoglobin has been clinically tested on humans, artificial blood will successfully and completely replace human blood.

So much for what and when to reference. But you also need a consistent and clear system for documentation of sources. Referencing is basically a two-part process: first, you cite your sources directly within the text; second, you include a complete reference list of all sources cited at the end of your paper.

As we have already mentioned, you need to reference not only direct quotes, but also anything at all that you paraphrase. To distinguish between direct and indirect referencing, use the following format:

> "Telomeres are specialized structures found at the natural ends of eukaryotic linear chromosomes" (Lundblad and Szostak, 1989).
> The quotation marks indicate that this sentence was taken directly from the source (word for word) without any alterations. Direct referencing is *not common* in scientific writing.

> Telomeres are located at the ends of chromosomes to ensure that the entire chromosome is copied properly (Lundblad and Szostak, 1989).
> The lack of quotation marks indicates that this idea was paraphrased from the source. You must always credit others' ideas, even if you do not quote them directly.

Referencing in the sciences, in contrast to the humanities, is not standardized. The particular referencing style depends on the document's purpose. In science writing, you usually cite peer-reviewed journal articles – not monographs or textbooks – because journal articles represent the most current source of information. There are two main scientific referencing styles, both of which can be further adapted to your specific needs or purposes. (Also, see "Choosing a Reference Style," p. 32.)

Expanded Referencing Style

This style places the authors' names and date of publication in the text and lists all the references at the end of the report in alphabetical order. Its advantage is that the author(s) is/are referenced right in the text; thus, those familiar with the field will immediately recognize the reference. The date is important so that the reader can distinguish which study is being cited (often more than one article by an author[s] is being referenced).

> More complex repeat sequences are also associated with the ends of chromosomes; in *S. cerevisiae*, two such repeated sequences are the Y and X elements (Chan and Tye, 1983).

Two rules apply when you are using this format. If there are only two authors, as in the example above, cite both names. If there are more than two authors, cite only the first author's name followed by "*et al.*" in italics with the date. *Et al.* is the abbreviated form of the Latin expression *et alii*, which means "and others."

> Mice with a mutation in the *fer* gene will form abdominal tumors within ten days of being exposed to a teratogen (Mezyk *et al.*, 2000).

At the end of your paper, always include a complete list (in alphabetical order) of all the references cited. If you list more than one work by an author(s), order them by date in the reference list (from most recent to oldest).

References

Bednarek, A., Budunova, I., Slaga, T.J., and Aldaz, C.M. (1995). Increased telomerase activity in mouse skin pre-malignant progression. Cancer Res. 55, 4566–4569.

Blasco, M.A., and Greider, C.W., (1996). Differential regulation of telomerase activity and telomerase RNA during multi-stage tumorigenesis. Nat. Genet. 12, 200–204.

Blasco, M.A., and Greider, C.W. (1995). Functional characterization and developmental regulation of mouse telomerase RNA. Science 269, 1267–1270.

The style you use may or may not require the title in the reference list. Punctuation may also vary; for example, the volume number may be bolded in some journals, or the date may sometimes be placed at the end of the citation. Journal names are abbreviated in a standard way. These abbreviation conventions can be found in the reference section of any library (see, for example, Council of Biology Editors, *Scientific Style and Format: The CBE Manual for Authors, Editors, and Publishers*, listed in appendix 1).

An example of a report using the expanded referencing style, in its body and a reference list, follows:

The Function of Imprinting in Mammalian Development
Shirley M. Tilghman

A question of central importance to the field is the functional significance of genomic imprinting in mammals. To date autosomal imprinting, as defined in this review as the differential expression of the two parental alleles of a gene, has been demonstrated only in eutherian (i.e., placental, nonmarsupial) mammals. However, female marsupials exhibit a form of genomic imprinting as they preferentially inactivate the paternal X chromosome in all somatic cells (Cooper et al., 1993). A similar mode of imprinted X inactivation occurs early in development in the extraembryonic tissues of eutherian females (Takagi and Sasaki, 1975) and may well represent the ancestral form of X inactivation.

The highly restricted developmental potentials of androgenotes with two paternal genomes, and gynogenotes or parthenogenotes with two maternal genomes, was interpreted to mean that genomic imprinting was critical to development in mammals. In retrospect this argument is difficult to sustain. Androgenotes and gynogenotes do not fail to imprint completely; rather they have a genome-wide imbalance in the dosage of imprinted genes.

For example, androgenotes will have double the dosage of paternally expressed genes and no expression of maternally expressed genes. The same condition holds for single or partial chromosomal uniparental disomies, which have been extensively studied in mice, and often lead to developmental anomalies (Cattanach, 1986). This leaves open the possibility that imprinting is dispensable under conditions where the imprints on both parental genomes are erased, as Jaenisch (1997), has suggested. There is some indirect evidence for this. Parthenogenetic embryos that are generated from nuclei of immature oocytes, that may be at a stage of development when imprints are not fully established, develop to a later stage than parthenogenotes from more mature oocytes whose imprints are in place (Kono et al., 1996).

(Taken from: Tilghman, S.M. (1999) Cell 96: 185–193. Copyright © 1999 by Cell Press)

References

Cattanach, B.M. (1986). Parental origin effects in mice. J. Embrol. Exp. Morph. Suppl. 97, 137–150.

Cooper, D.W., Johnston, P.G., Watson, J.M., and Graves, J.A.M. (1993). X-inactivation in marsurpials and monotremes. Semin. Dev. Biol. 4, 117–128.

Jaenisch, R. (1997). DNA methylation and imprinting: why bother? Trends Genet. 13, 323–329.

Kono, T., Obata, Y., Yoshimzu, T., Nakahara, T., and Carroll, J. (1996). Epigenetic modifications during oocyte growth correlates with extended parthenogenetic development in the mouse. Nat. Genet. 13, 91–94.

Takagi, N., and Sasaki, M. (1975). Preferential inactivation of the paternally derived X chromosome in the extraembryonic membranes of the mouse. Nature 256, 640–642.

Abbreviated Referencing Style

In this style, numbers are placed parenthetically within the text and all the references cited are listed at the end of the article in numerical order (i.e., the order in which they were cited in the paper). The advantage of this style is that it is shorter and does not clutter the text with additional words and numbers (neither names nor dates are included within the text itself).

> More complex repeat sequences are also associated with the ends of chromosomes; in *S. cerevisiae,* two such repeated sequences are the Y and X elements (3).

If the reader wants to know this particular source, he/she may look up no. 3 in the reference list. In some journals, the number is in parentheses; in others, it is in superscript form. If you list more than one work by an author(s) in your numbered list at the end of your article, each citation has a different number and date. Following is a sample list of references that follow the abbreviated referencing style.

References

1. Rittling, S. (1996). Exp. Cell Res **229**, 7–13.
2. Bednarek, A., Budunova, I., Slaga, T.J., and Aldaz, C.M. (1995). Cancer Res **55**, 4566–4569.
3. Chan, C.S.M., and Tye, B.K. (1983). Cell **33**, 563–573.

As a space-saving measure, this style may not include the title in the reference list. Again, punctuation in this style may also vary.

Choosing a Reference Style

First and foremost, choose one style and adhere to it strictly. When in doubt, ask your professor or instructor whether they have a preference. If they do not, select a journal and follow its style and format. Be sure to include all the proper punctuation and formatting!

Argon-Lead Isotopic Correlation in Mid-Atlantic Ridge Basalts
Philippe Sarda, Manuel Moreira, Thomas Staudacher

Rare gas isotopes have been successfully used to constrain the timing of atmosphere generation by mantle degassing (1–5), but, except for helium (6, 7), they have been less useful as tracers of mantle heterogeneities, largely because of contamination by atmospheric rare gases (8, 9). Measuring the isotopic composition of mantle rare gases trapped in mid-ocean ridge basalt (MORB) has always met the difficulty of separating the magmatic gas from a widespread atmospheric component, a particular problem for argon. Addition of atmospheric Ar was long considered to occur during eruption on the sea floor. Indeed, the slowly cooled inner parts of submarine lava flows generally appear more contaminated than the outer layers, quenched to glass (8, 9). In step-heating degassing on glass samples, argon with $^{40}Ar/^{36}Ar$ ratios of 300 to 5000 is often released in the low-temperature steps (500° to 700°C). This gas is confirmation that weakly bound atmospheric argon is present in many samples (9–13).

(Reprinted with permission from Sarda, P. *et al*. (1999). Science 283: 666–668. Copyright 1999 American Association for the Advancement of Science.)

References

1. F. P. Fanale, Chem. Geol. 8, 79 (1971).
2. M. Ozima, Geochim. Cosmochim. Acta 39, 1127 (1975).
3. C. J. Allègre, T. Staudacher, P. Sarda, M. D. Kurz, Nature 303, 762 (1983).
4. C. J. Allègre, T. Staudacher, P. Sarda, Earth Planet. Sci. Lett. 81, 127 (1986).
5. J. Kunz, T. Staudacher, C. J. Allègre, Science 280, 877 (1998).
6. M. D. Kurz, W. J. Jenkins, J.-G. Schilling, S. R. Hart, Earth Planet. Sci. Lett. 58, 1 (1982).
7. J. Dymond and L. Hogan, Earth Planet. Sci. Lett. 38, 117 (1978).
8. P. Sarda, T. Staudacher, C. J. Allègre, Earth Planet. Sci. Lett. 72, 357 (1985).
9. D. E. Fisher, Nature 256, 113 (1975) .
10. T. Staudacher and C. J. Allègre, Earth Planet. Sci. Lett. 60, 389 (1982).
11. M. Moreira, T. Staudacher, P. Sarda, J.-G. Schilling, C. J. Allègre, Earth Planet. Sci. Lett. 133, 367 (1995).

Referencing Other Types of Sources

1. Textbooks

Although referencing textbooks is not common in science writing, you may need to cite your textbook in a lab report. The following represents one way to reference a textbook:

Lodish, H., Baltimore, D., Berk, A., Zipursky, S.L., Matsudaira, P., Darnell, J. (1995). *Molecular Cell Biology* 3rd ed. W.H. Freeman and Company, New York, pp 24–30.

W.H. Freeman and Company is the publisher; New York is the place of publication; and pp indicates what pages were referenced.

2. Book Chapters

Compiled volumes are referenced slightly differently than textbooks are. In this case, each chapter is written by a different author.

Koornneef, M., and Stam, P. (1992). Genetic Analysis. In *Methods in Arabidopsis Research*. Eds. Koncz, C., Chua, N-H, and Schell, J. World Scientific Publishing Co. Ltd., Singapore, pp 83–99.

3. Web Sites

A newer source of information is, of course, the internet; all website sources that you consult must be referenced. The appropriate style, according to the CBE Manual published by the Council of Biology Editors, lists the author's name (if available), date of internet publication, document title, URL, and date of access.

> Brians, P., February 1997, Common Errors in English, www.wsu.edu/~brians/errors.html, accessed: July 1999.

A note of caution: websites on the internet may or may not be peer-reviewed. This means that the information may not always be accurate. Be sure to cite only reliable websites!

How to Assess a Web Site's Reliability

Web sites vary wildly in terms of accuracy; no precise guidelines or check-lists can ensure completely that you are referencing a reliable web site. Paula Hammett (1999), however, argues that one should consider five main criteria when assessing a web site's reliability:

1. *Identification of Author* – Is the author clearly identified? Are his/her credentials given? Is the author associated with an organization or university? Is there a link from this person's web page back to his/her organization or university home page?

2. *Purpose* – What is the purpose of the web page? Is the intended audience clear? Does the page adequately cover the subject or is it simply an overview?

3. *Accuracy* – Does the author cite references for any facts that he/she provides, and is there a bibliography? From your own knowledge, does most of the information seem accurate? Does the web page appear to be sloppily written, or does it have spelling errors? Such carelessness indicates that the author paid little attention to the details of setting up the web page and reflects badly on the site's reliability.

4. *Date* – Can you determine when the web page was initially placed on line? Has it been updated recently?

5. *Integrity of Information* – Does the web page rely solely (or too much) on photographic material to make its point? If so, be careful; photographs and digital images can be easily modified.

(Hammett, P. (1999). Teaching tools for evaluating world wide web resources. Teaching Sociology 27:31–37.)

Chapter Two

Other Common Types of Scientific Writing

PREPARING A POSTER PRESENTATION

The primary purpose of a poster presentation is to illustrate the main research question(s) of a particular project, as well as the results you have obtained thus far. Most commonly used at a scientific conference, a poster gives conference participants the direct opportunity to acquaint themselves firsthand with current research in their field. Generally, several posters are hung in one room, where people can wander from poster to poster, stopping to view those that interest them the most. The poster's author, usually present beside his/her poster, fills in any details that he/she has not been able to include on the poster and answers any pertinent questions. Thus, through this visual and oral medium, scientists are able to learn about new research ideas, approaches, and directions from their peers without having to listen to (or give) a formal presentation. Since the author ordinarily accompanies his/her poster (and presents its contents informally), its text should be brief and concise, habitually formatted in a relatively large font (20 points), with 1.5 line spacing.

Title: The title is in an extremely large (40–60 point) font and includes the authors' names and addresses. E-mail addresses are also particularly useful.

Introduction: This part provides the relevant background (in the first paragraph) and clearly indicates the purpose of the study. It is usually no longer than two pages, in a large (20 point) font.

Materials and Methods: This section may or may not be present. Since poster space is often limited, *Materials and Methods* tends to be the first section omitted. If it is included, it is very brief, including only those protocols that are unique in that particular field.

Results: This section typically lacks any text at all. Instead, figures, along with detailed figure legends, effectively present the preliminary results of the study. A graphic, non-textual format simplifies access for casual viewers; by simply glancing at your poster, someone can easily see and interpret your data.

Discussion: This part can be in either sentence or point form. The advantages of point form are that it is easy to read and it emphasizes the results' relevance to the overall study. Include your conclusion at the end of this section.

References and Acknowledgments: References and acknowledgments may or may not be present depending on space restrictions. If they are present, they should be few in number.

Poster Layout: More often than not, authors write posters from top to bottom and then proceed from left to right because it is easier for readers to follow a large poster bulletin vertically in columns (as they would read a newspaper) than it is for them to follow it all the way across horizontally. In addition, such a layout permits several people to read the poster concurrently.

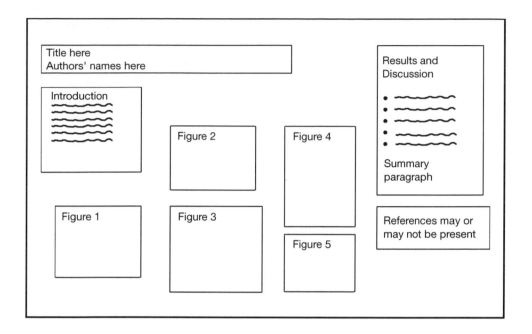

WRITING A PROPOSAL

Proposals have several different formats and may be used for various purposes. The most common type of proposal is the research **grant proposal** that senior graduate students and professors have to write in order to obtain funding for a particular project. In this case, the author(s) must "sell" the significance and quality of the research proposed. Since most funding agencies have many qualified applicants, proposals must be persuasive in terms of purpose and rationale. Essentially, when you write a grant proposal, you must sell both yourself and your project to the funding agency.

A second type of proposal, a more common one for students, is the **research proposal**. Often, fourth-year undergraduate, as well as graduate, students involved in research projects are required to write research proposals at the projects' outset. A research proposal serves two purposes: first, it allows the students to complete all the necessary background reading and to place the research question in context; second, it outlines the experiments (often with a time line) that will be performed to answer this question.

Thus, by writing a research proposal, a student demonstrates that he/ she is familiar with the relevant background material, has formulated a hypothesis or focus statement (chapter 1), and has developed a directed approach to answering the question. Some research proposals contain a list of objectives, as well as a feasible time line in which the researcher expects to achieve these objectives.

A research proposal should:

- introduce the topic and present relevant background information;
- indicate how the proposed research fits in with what is already known in the field;
- use the literature to support its intended directions;
- clearly state the hypothesis;
- address the who, what, where, and why early on;
- outline the rationale for answering the question;
- describe the experiments needed to answer the question;
- summarize the expected results of the experiments, if possible;
- include appropriate and correct references;
- conclude, in the final section, why the research is relevant and important; and
- be proofread by a colleague to eliminate errors.

If the format permits you to do so, structure your proposal in an interesting and readable way. Use headings to guide the reviewer and stay focused. Try to pique the reviewer's interest in your proposed work (tell him/her why he/she should care about it) and make sure to balance short- and long-term goals. Too commonly, proposal writers present an abundance of short-term goals, neglecting the long-term ones. Researchers too often overlook the bigger picture, since they tend to be preoccupied with the minute details of the research process. So, be sure at all times to keep the big picture in mind and communicate this to your readers.

Find out about the exact deadline for, and precise form of, the proposal's submission. Many proposal forms are now available on-line and can be submitted electronically. If the work can be submitted electronically, verify that the due date is midnight of that day (you may need the extra hours!).

Finally, look over your proposal and decide the following:

- Does the structure of the proposal help the reader?
- Do the tables and/or figures present the data clearly?
- Is the intent obvious?
- Is the writing style straightforward?
- Is it clear how the hypothesis or focus statement relates to the literature?
- Is the proposal properly referenced and have you acknowledged the appropriate people?
- If you read this proposal as a reviewer, would you be convinced of its importance?

GETTING STARTED ON A *RESEARCH PROPOSAL*

Prepare the title and introduction as you would for a lab report. In your introduction, briefly summarize all current background knowledge necessary to understand the significance of your research question. Use proper references and clearly state the research question or hypothesis at the end of your introduction. Some research proposals have clearly defined sections, while others do not. Acquaint yourself with the format expected in your case and use it appropriately for your purpose. In your introduction, present a clear and concise vision of your research question and of how it relates to the bigger picture as described in the previous research.

Next, explain your rationale or experimental design. Describe the purpose of each of your experiments, as well as which tools will be used (e.g., PCR, northern analysis). Then, outline the expected results for each of your experiments and discuss the potential significance of such results. If you do not know (or cannot predict) the outcome of a particular experiment, then hypothesize about possible outcomes. If you have any preliminary results, insert them in this section. Point out that future work will build on these results.

A note of caution: Some proposals focus primarily on future work and contain a brief introduction, while others emphasize a well-developed introduction, thus downplaying proposed research to a certain extent. Find out which structure is appropriate in your particular case and use it.

Chapter Three

The Academic Research Essay

AN OVERVIEW

The English word "essay" derives from the French *essai*, meaning "attempt." Michel de Montaigne first used the term in 1580 to describe personal, exploratory writing about daily life, but the term "essay" has evolved over the centuries to include formal writing on a wide range of subjects. Meanwhile, the root and prefix meaning of the word "research" is "to seek out again." Most of the research you do for university essays will entail seeking out the ideas and materials of others, usually in order to analyze and comment on these in new (and often untried) ways.

A research essay differs from a report. For an essay, you need to **evaluate, interpret,** or **analyze** – or in some other way add to, and participate in – what you read and write about. You are expected to consider the "why" and "how" of the topic you are writing on – you should, in the process, be guided by a key, intriguing question or a set of such questions (it is not enough simply to report the ideas of others in a research essay!). Reports also involve critical thinking; in general, however, they emphasize description and explanation more than do essays. In addition, unlike essays, reports are structured somewhat rigidly by headings and/or subheadings.

In an essay, you have to develop a focused point of view (or approach) towards your material. This perspective, which must be based on solid evidence and logical thought, helps you to form your argument (or **thesis**). It is usually a partial (or sometimes more than partial) answer (or answers) to your particular question (or set of questions). But it is ALWAYS an answer that is based on serious and substantial research. This thesis, once you have determined and articulated it, represents the condensed kernel of your whole essay: it defines its purpose and gives it **unity** and **coherence**. For this reason, the thesis is usually (but not always) stated explicitly in your introduction (usually near its end).

Let us summarize briefly what we have said so far by indicating what a research essay *is*, as well as what it *is not*.

A research essay IS
- a synthesis of your discoveries about a topic, as well as your interpretations and evaluations of those discoveries;
- a work that demonstrates some originality; and
- a piece of writing that acknowledges all the sources to which you are indebted.

A research essay IS NOT
- a summary of the articles, books, or other source material that you have consulted;
- an uncritical repetition of the ideas of others;
- a series of quotations, arranged together in a clever or skillful way;
- unsubstantiated, personal opinion that lacks supporting evidence; or
- the wholesale copying of the ideas or findings of other scholars or writers (which is "plagiarism," considered a serious academic offense [see chapter 1: References, p. 26]).

The final product (the essay itself) does not write itself quickly or automatically – within a few hours or overnight. Instead, it is the last stage of a complex and lengthy process of thinking and rethinking, then of writing and rewriting. We will break this process down into what we consider logical stages. Experts on writing have found that an effective writing process is made up of a series of stages that are interconnected. However, there is no one "right way" to write an essay; the more *you* write, the more *you* will discover what works best for *you*. Try to view each essay assignment as a valuable experience to learn research, as well as critical reading, thinking, and writing skills!

STAGES IN THE PROCESS

The main stages in writing an academic research essay are shown in the following box:

Prewriting: Focusing, Researching, Planning
- A. *Choosing and/or Clarifying your Topic*
- B. *Gathering Information*
    - 1. Making a working bibliography
    - 2. Taking notes
    - 3. Organizing your information and ideas
- C. *Developing a Thesis*
- D. *Writing an Outline*

Writing the First Draft
- A. *Drafting the Body*
- B. *Composing an Effective Introduction*
- C. *Writing a Successful Conclusion*

Redrafting/Revising and Editing
- A. *Revising*
- B. *Editing/Proofreading*

Referencing
- A. *Preparing the Final List of Works Cited*
- B. *Preparing Final Source Citations*

Let us deal with each of these stages separately, albeit briefly.

Prewriting: Focusing, Researching, Planning

This stage is really the most important of all; therefore, we will devote the most time to it (as *you* should!). Beginning writers seldom realize that the most important part of writing occurs even before they begin the first draft. This prewriting stage involves a whole series of interrelated steps.

Choosing and/or Clarifying the Topic

The first step, *always*, is choosing or (if you have already been handed one) clarifying your topic. If you have to choose your own topic, then it should satisfy certain basic criteria (otherwise, you will experience great difficulties researching and writing the essay, and the final product will be flawed). Here are these basic criteria. A topic should

1. contain a significant or controversial question (or questions) that you will focus on and seek to answer in your essay;
2. interest or captivate you;
3. be of manageable scope (not too broad); and
4. be appropriate for your audience.

You can develop topics using a variety of methods, including focused free-writing, making lists, and asking probing questions about the subject. You can find out more about these techniques by consulting the books in appendix 1, or by visiting the writing center or writing lab in your college or university and meeting with a writing tutor.

If, however, you have already been given a specific topic by your instructor (e.g., "Examine the assaying of DNA as a tool for protecting endangered species" or "Outline and evaluate the mathematical contributions of Leibniz"), then your task is somewhat simpler. However, before you begin any research, you will still have to clarify your topic, reflecting carefully on its meaning so that you understand as clearly as possible what your professor or instructor is asking and/or expecting of you. First of all, note the key terms in the question. Are you being asked to **"analyze"** (look behind the surface structure and examine the parts in detail, trying to determine the relationship of these parts to the whole and asking questions like "how" and "why"), to **"compare"** (examine similarities *and* differences in order to understand the objects or ideas more clearly), or to **"evaluate"** (determine the strengths or weaknesses of something [e.g., a theory, a methodology] by applying your judgment to the results of your analysis)? Terms such as these are extremely important; pay attention to them! Also, when clarifying your topic, note whether you are being asked to employ any particular concepts or methods. (Does the question, for example, require you to evaluate a theory by applying it to an example from outside the course material?)

Gathering Information

Now you are ready to begin researching your question, that is, collecting and gathering information that will help you answer the question. Be as systematic as possible when you begin this process in order to avoid becoming lost in a mass of information and to stay "on track." You might want to develop a schedule that breaks down the time you have available for the necessary work (you can use the outline provided in this chapter as a template for such a schedule).

Whatever your topic, you have access to a huge variety of sources – both print and electronic. These include reference books (encyclopedias, dictionaries, atlases, etc.), periodicals (magazines, journals, and newspapers), books and monographs, and electronic sources (CD-ROM, internet web sites, computerized databases). Personal interviews (conducted by you) are one other source of information, often useful for essays in certain social science or humanities courses.

1. Keep track of your sources by compiling a **working bibliography** – a file of the books, articles, and other sources you think will be useful. When your file seems complete, decide which sources seem the most useful and consult them first. If possible, put them on file cards and include all pertinent information (title, author, publisher, place and date of publication).

2. Begin **reading and taking notes** from the sources when your working bibliography appears to have covered the key aspects of your topic. For a paper of 2000 to 2500 words, ten to fifteen promising titles should give you a good start. However, be sure to check with your instructor on how many sources are expected!

It is essential that you take detailed and clear notes from your sources. While doing your research reading, keep your question clearly in mind – have it written out in front of you; you will thus read with a more specific purpose and be able to determine which material is the most relevant. Look for facts and theories to help you answer your question, as well as the opinions of others on whether certain answers seem logical. In the process, attempt to formulate a provisional thesis or hypothesis. You do not have to be totally committed to this hypothesis, but pursue it as you read more, testing it as you go along.

Do not write down too much. Instead, note only those ideas that are relevant to your focus on the topic; also, try to **summarize** as much as possible, as opposed to **paraphrasing** or **direct quoting.** Summarizing means condensing an extended idea or argument into a sentence or more in your own words. Sometimes, however, paraphrases or direct quotes are helpful when you want either (1) to reconstruct an author's line of reasoning or (2) to include the unique effect of the author's own words. In addition, do not depend on highlighting and underlining! (This method of note-taking will not force you to think through and test your own understanding of the idea.) Leave space on your note cards for comments of your own, as well as for questions and reactions that might be added over time.

Be sure to label your notes well. Take notes on separate file cards and then label the topic of each card. Then, you can classify and/or synthesize your ideas in different ways later on. Record bibliographic information on a master list *before* you begin note-taking so that you can quickly and concisely include the source on each note card.

3. How do you know when to stop taking notes and begin **organizing** your thoughts and information? After you have examined all your major sources, identified important connections among your notes, and developed some ideas of your own about the topic, you are probably ready. Once you can see beyond each individual note to the beginning of a larger cohesive argument, you should have a more complete sense of your work. Now you can start reviewing and sorting all your notes.

Read each card, deciding which particular aspect of your topic it deals with and group your cards according to these categories. Some aspects may seem less important than they did at the beginning of your research; you may even discard some of your notes at this stage. Equally, you may even discover areas that require further investigation.

Developing a Thesis

Now is the time to start "massaging" your provisional thesis – "kneading" it and working it through more thoroughly so that it becomes a specific and unambiguous statement (or series of statements) about your topic. A good thesis statement establishes a kind of contract between you – the writer – and your reader. It should indicate a precise subject (e.g., "The two most common methods of cloning DNA"), as well as a definite and limited assertion about this subject that needs to be explained and supported by further discussion ("Of the two most common methods of cloning DNA, i.e. PCR and the use of plasmid libraries, PCR is the more efficient"). Thesis statements may also show the ways in which you will use evidence to argue your point and even your awareness of possible disagreements or counter-arguments, as well as ways to dispute these.

We think it important to underline what a thesis statement is *not* and, in so doing, perhaps counter some prominent myths about this element. A thesis statement is NOT

- a promise or statement of purpose (e.g., "In this essay, I will demonstrate how mitochondrial DNA analysis is used in conservation genetics.");
- a topic or subject only ("New Drugs to Stop the Replication of the Influenza Virus") (this is more a title than a thesis statement);
- a question ("How do the rates of ATPase activity of three plant membranes – the mitochondrial membrane, the tonoplast, and the plasma membrane – compare?");
- *always* the last sentence of the first paragraph (This is often a natural place for it, but it is not the only one. Sometimes, your opening sentence may include your thesis; other times, you may not fully formulate your thesis until the end of the essay);

- *always* only one sentence in length (Many thesis statements require two or three sentences or even entire paragraphs in order to be clearly and fully articulated); or
- *always* indicative of three supporting points (It will contain some evidence for its assertion, but no magic rule stipulates how many pieces of evidence are needed).

Writing an Outline

Once you have articulated a relatively precise and well-defined thesis statement with which you are satisfied, you can begin your outline. An outline is an orderly plan showing the division and arrangement of your ideas – in other words, the framework according to which you will "build" or organize the development of your argument, the essay itself.

Many of you skip this stage at your own peril. Even though you may have a structure for the paper in your mind after writing down your thesis statement, the actual writing of a research paper usually takes a great deal of time. So it is easy to forget things later that you thought you had firmly in your mind when you began. An outline, in this sense, serves as a handy reminder of your key ideas and the relationships among them. Second, an outline allows you to evaluate the effectiveness of your organizational plan and rearrange ideas easily at an early stage. Once you have written something in paragraph form, rearrangement often becomes difficult and time-consuming.

How do you go about constructing an outline? What kinds of outlines are there? Informal outlines follow any pattern that a particular writer finds helpful, but formal outlines fall into three categories: the paragraph outline, the topic outline, and the sentence outline. The **paragraph outline** is a list of numbered sentences, each one summarizing subsequent paragraphs in your essay. Such an outline helps you review the progression of ideas, but it does not reveal the actual development of each paragraph (the kinds of evidence you will use) or the logical relationships between and among the various paragraphs. For this, you are better to use the topic or sentence outline.

Begin a **topic outline** with your entire thesis statement. Then, write brief phrases on separate lines, numbering and lettering them to show their relationship to one another. Generally, the structure resembles the following (a sample, somewhat abridged outline):

Thesis: Despite some possible drawbacks, Genetically Modified (GM) crops are probably, on the whole, beneficial to human health and the environment. Nevertheless, government policy needs to be more clearly defined and applied in this area.

I.   Genetically modified crops may be harmful
   A.   Short-term and long-term effects are unknown
      1.   monarch butterfly example with GM bt corn
      2.   rice example with added vitamins
   B.   Novel genes may spread to natural populations
      1.   antibiotic example
      2.   monarch butterfly example
   C.   Human allergies to modified proteins
      1.   brazil nut example in soybean
      2.   pollen proteins in wheat
II. Genetically modified crops have many benefits
   A.   Facilitate farmers
      1.   farmers have to use fewer herbicides (cost/labor)
      2.   cotton example
   B.   Developing world could produce more crops for their people
      1.   tomatoes are engineered to ripen only when induced (no refrigeration needed in transport)
      2.   fewer chemical sprays needed to maintain crops
III. Government Policy on Genetically Modified Crops
   A.   Who will regulate this?
      1.   government
      2.   companies
      3.   overseeing agency
   B.   Which aspects of GM crops should be regulated?

As you can see, the essay's main arguments are identified by Roman numerals; each of these ideas should be roughly equivalent to one another in terms of logical significance. The supporting sub-ideas (or sub-arguments) under these main ones are then identified by capital letters. Again, these should also be roughly equivalent in terms of importance. The next level is identified by Arabic numbers. Note that all headings and subheadings are "content rich." Do not use vague, useless phrases like "introduction," "main body," "examples," or "reasons" in your outline.

Writing the First Draft

Okay. The research is finished, you have come up with what you think is a pretty good thesis, your outline is more or less completed – and you think you are ready to write. So why is it so difficult to get going? Why do we all find a million excuses – umpteen other things that need doing as that deadline looms? One reason is that the whole process of drafting your essay is a very uncertain one. We do not transcribe fully formed and polished thoughts into words, sentences, and paragraphs automatically; instead, we often struggle to find and convey the meaning of our thoughts through the act of writing. Each of us combines the process of drafting and revising differently: some of us draft and revise at the same time, trying to perfect each unit before going on to the next. Others write madly on and on until all their thoughts are down on paper, rarely or never stopping to reread and rewrite until a first draft is complete. Attempt to find the method that suits your particular style of thinking – and writing – the best.

Here are some tips to "get going" initially – to cure your "writer's block":

- Write a paragraph on what you think your essay will be about when you finish it.
- Conjure up an image that represents your topic and describe that image.
- Read what you've already written – notes, outlines, etc. – and immediately start free-writing about whatever comes to mind.

- On the basis of your outline, divide your essay into parts. Start writing the part that seems most ready to be written – the one you understand the best or feel the most strongly about.
- Skip the opening and start with the body. Or write the conclusion first!

A. Drafting the Body

In the body of the essay you will explain, support, and develop your thesis. Remember that you are trying to do more than just string quotations together; instead, you want to produce original and coherent writing that supports and develops your argument.

As you draft the body, you should always keep both your purpose(s) and audience firmly in mind. Is (are) your purpose(s) to inform or persuade? Knowing your main purpose(s) will help you to establish the structure and style of the paper. Second, how do you characterize your audience (your professor or instructor? other students? the scientific community at large? a lay audience?)? Further, what do you need to be sensitive to in your audience's background (age, level of education, familiarity with topic)? What is your audience's attitude towards your topic? The type of audience you are addressing will determine, to a great extent, your content and tone.

The following are some key principles of effective writing.

Unity and coherence: Stick to your topic and thesis throughout. Be careful not to go off on tangents as you write. Coherent writing hangs together well and moves readers easily from one point to another. A well-structured outline helps; other techniques to achieve this end include the following:

- using transitional words and phrases to signal movement to your readers:
  - to show addition: and, also, besides, further, furthermore, in addition, moreover, next, too, first, second
  - to give examples: for example, for instance, to illustrate, in fact, specifically
  - to compare: also, in the same manner, similarly, likewise, equally

- to contrast: but, however, on the other hand, in contrast, nevertheless, still, even though, on the contrary, yet, although
- to summarize or conclude: in other words, in short, in summary, in conclusion, to sum up
- to indicate logical relationship: if, so, therefore, consequently, thus, as a result, for this reason, since, hence
- employing pronouns with clear antecedents

    The *first hemoglobin solution* was manufactured in 1937. When *it* was tested on animals, *it* was found to be "effective in the delivery of oxygen, but was highly toxic to the kidney" (Chen 1999).

- repeating key words and phrases (not in precisely the same wording)
- maintaining a consistent point of view
- integrating your evidence carefully and skillfully

Sufficient support: Support for the various assertions and arguments in your paper takes various forms – statistics, examples, authoritative opinions, documentation, etc. The more specific details and reliable sources you have, the more persuasive your argument will be.

Emphasis: In general, the more space you give to a particular idea, the more significant it will seem to your reader. So plan to develop points that are approximately equivalent in terms of their importance with equal amounts of evidence and discussion. Both repetition and strategic positioning of key ideas also help to achieve emphasis. Repetition means that you present an idea at least twice, each time with altered wording. But use this technique sparingly! Strategic positioning entails placing an idea in a featured spot: arrange your arguments (or pieces of evidence), for example, in order of increasing importance. The steady accumulation of proof may convince a skeptical reader.

Concreteness and specificity: When possible, use concrete and specific words and phrases instead of abstract or general ones.

B. Effective Introductions
Many writers choose to write their introductions last or second last (just

before doing their conclusions). This makes sense, for it is often easier to write your introduction once you know where your paper is actually heading. An effective introduction is crucial – it should focus your readers' attention on the topic and arouse their curiosity about what you are going to say in your essay. It should also be concise and move, as effectively as possible, from the general towards the specific. One of the most "tried and true" methods of constructing an introduction is to open with a broad statement of the general topic and then clarify or limit this topic in the next few sentences, finally articulating the thesis statement, which will assert the focus of the essay, towards the end. But many other techniques exist, including the following:

- using a quotation to lead into a thesis statement
- relating an incident or creating an image as a preamble to the thesis
- posing a controversial question or a startling opinion
- making a historical comparison or contrast (when some background to the essay topic is useful)

Strategies to avoid include the following:

- DO NOT rely on vague generalities and/or repetition before moving to your thesis to "get things going."
- DO NOT start with "The purpose of this essay is ..." or any other flat announcement of your intention.
- DO NOT refer to the title of the essay in the first sentence: "The above title poses an interesting dilemma ..."
- DO NOT begin with "According to the Oxford Dictionary ..." or any kind of dictionary definition. This opening is too clichéd.
- DO NOT apologize for either your opinion or inadequate knowledge of the topic.

C. Successful Conclusions

Your conclusion is not simply a signal to your readers that you have stopped writing; it is, effectively, an assurance that the essay is completed – that it has been brought to its appropriate climax. Thus, a conclusion should be more than a sentence or paragraph attached to the end of the essay with a mechanical transition like "Therefore ..." or "In conclusion ..." Instead, a successful conclusion "ties everything together" or summarizes the evidence presented in the essay, but it usually does much more than that. For example, it could

- restate the thesis with a fresh emphasis;
- remind the readers of what you want them to do or think about the topic (suggest a course of action);
- pose a significant question that may be implied by your thesis;
- link what you have written either to something known or to what seems a future possibility; or
- make connections between/among ideas explored in the paper.

As you can see, the possibilities really are endless. Be creative, but give your readers a strong sense of closure!

Certain techniques should, however, be avoided. For example:

- DO NOT simply reiterate your introduction. A conclusion that is nearly identical to the introduction indicates a weakly developed essay.
- DO NOT bring up new topics, either ones broader than those your essay has dealt with or ones that need extensive clarification.
- DO NOT apologize for your argument or cast any doubt on it. ("Even though I am not an expert, I conclude that ...")
- DO NOT repeat explicitly what you have already stated in the paper ("In this paper, it has been shown that ...")
- DO NOT stop abruptly or simply trail off.

Redrafting: Revising and Editing

Redrafting is an integral part of the writing process. Writing, in fact, largely consists of redrafting, even though the final draft reveals little of the actual revision process. Many of us, however, neglect this extremely important stage, perhaps in part because of a common misconception that a writer cannot or should not redraft his/her own work. All too frequently, student writers simply skim their first draft quickly for glaring errors just before the assignment is due, and then submit it immediately. Experienced writers revise and redraft their work at least two to three times.

A. Revising

Revising is not the same thing as editing (or proofreading). Revising generally entails "re-envisioning" your entire draft – that is, taking a fresh look at its argument, logic, organization, tone, style, etc. In other words, revision involves paying careful attention to the meaning that you want your paper to convey. You may even find yourself rethinking your aims and methods. To revise, you must evaluate your essay and ascertain exactly where improvements are needed, then determine how to implement them. Remember, you are trying to look at your writing from the perspective of your intended reader. So, first and foremost, you need to distance yourself from each draft and read it objectively, not as an overly involved writer does – giving yourself a "cooling off" period of at least a day or two (if not a week!)

Essentially, the revision process includes four main activities: (1) adding material or inserting necessary words, sentences, phrases, and paragraphs; (2) cutting anything that is off-topic or repetitive; (3) replacing inadequate words, phrases, sentences, or paragraphs with new ones; and (4) shifting material that does not seem to be presented in logical order. Sometimes, the hardest thing to do is to cut material because of authorial pride or a sense that whatever you struggled to put down in the first draft "deserves" to stay. Resist such feelings!

Here are some suggested strategies for revision. First you need to acquire a sense of the *whole* piece of writing. Read your entire paper out loud and slowly, alerting yourself to both strengths and weaknesses. Or, perhaps listen while someone else reads and responds to it. Peer response, or constructive criticism of your draft by a fellow student, can also be an excellent aid to revision. Use all your resources – your powers of logic and your intuition – to evaluate your work. What follows is a checklist of questions that may help you:

- How well does my essay conform with my original purpose? Is my reason for writing the essay clear to the reader, not only in my thesis statement but also throughout the essay? If the body of the essay does *not* carry out the thesis, is it because my thesis is improperly worded or because the body itself rambles?
- Does the introduction engage and focus the readers' attention? Does it lead effectively and clearly into the body?
- Is the essay unified? Does each paragraph relate clearly to the thesis?
- Is the essay coherent? Are the relationships within and among its parts apparent? Do the transitions take the reader clearly from one paragraph to another?
- Is the overall shape or organization of the essay obvious to my readers? Is this organization helpful or distracting?
- Is the overall balance among sections effective?
- Is the development of the argument sound and logical? Does it avoid logical fallacies and inconsistencies? Do the reasons and evidence support the thesis statement? Are generalizations supported by specific details and concrete facts? Which details work best? Which points need further support? Have any important facts/ideas been omitted?
- Are opposing positions mentioned and responded to?
- Does the essay raise any questions that it does not answer? If so, what are they?
- Do the tone and style of the essay convey an appropriate attitude towards the topic? Are the tone, register (informal vs. formal), and style appropriate for my purpose and audience? Are they also consistent throughout the essay?
- What metaphors or similes are used that illuminate the topic? Do any need further development?
- Is each paragraph in the body (a) unified, (b) coherent, and (c) well developed? (a) Do all the sentences relate to the paragraph's main idea? (b) Is the overall structure of the paragraph coherent and logical? (c) Is the paragraph's idea developed in some depth, i.e., is there sufficient evidence to support and elaborate on the main idea?

- Are the references complete and accurate? Have they been integrated properly?
- Does the essay's title hint at the scope and approach of the paper?
- Does the conclusion provide a satisfactory sense of completion?
- Are the headings, if any, appropriately placed and clear?

Obviously, there are many aspects to consider here. Try to deal with each of the above questions individually. You will likely have to go over the entire essay several times to make all needed revisions.

B. Editing

Editing is more a micro-process that involves looking for and correcting sentence-level errors and infelicities. The editing process can begin once you have done your major revisions, or you may want to do it simultaneously. Nevertheless, we strongly suggest the former strategy.

First, check for spelling errors. Although most of us use word processors with spell checkers, spelling errors still occur. For one thing, many scientific or specialized terms and phrases are not found in spell checkers, so you may find yourself pressing the "ignore" button too easily. To make the most of this feature, enter specialized terms that you use frequently right into your word processor's dictionary. Make sure you spell these words *correctly* in your dictionary and check them periodically!

Next, begin tackling the grammar and style errors or weaknesses. The following checklist of questions may aid you in the editing/proofreading process (more precise information on each of these aspects may be found in chapter 5):

- Have you eliminated all sentence fragments, comma splices, and run-on sentences?
- Have you eliminated confusing shifts in verb tense, voice, and mood?
- Have you eliminated misplaced and danging modifiers?
- Do all verbs agree with their subjects? (plural – singular)
- Do all pronouns (it, them, its, their, etc.) agree with their antecedents?
- Is all your punctuation correct and complete?

- Are your sentences concise?
- Do your sentences show clear relationships among ideas?
- Do you use parallelism effectively and properly?
- Does your wording display variety and proper emphasis?
- Have you used precise and appropriate words?
- Do your words reflect an appropriate level of formality?
- Have you avoided sexist language, slang, clichés, artificial language, and unnecessary jargon?

A thoroughly revised and well-edited paper is well worth the effort. It is easier (and more enjoyable) to mark as well! Asking a peer or colleague to have one final look at your paper is also an effective strategy: professional academics do this all the time. We have included below some standard editing notations and symbols. You may find them useful, or, alternatively, you may want to come up with your own.

EDITING NOTATIONS:

agr	agreement of subject and verb	sp inf	split infinitive
amb	ambiguity	T	tense
awk	awkward	trans	transition
cap	capitalize	wdy	wordy
cs	comma splice	ww	wrong word
D	diction	‖	Parallelism
Dang	dangling modifier (or dm)	....	Undo deletion of word or letter
exp	explain	§	New section
frag	sentence fragment	~~~	Bold word or letter
gr	grammar	—	Italicize word or letter
mod	misplaced modifier	...	Undo deletion of word or letter
P	pronoun reference	^	Insert text
quot	quotation marks	¶	New paragraph
run on	run-on sentence	∽	Change order of words
ss	sentence structure	ℓ	Remove letter or word from text
sp	spelling	⌣	Join words together

Referencing

A. Preparing the Final List of Works Cited

After completing the revising/editing process, prepare a final list of all the sources you used in writing your essay. Include all those you have quoted, paraphrased, or summarized. Usually, this list is entitled "Works Cited." It is generally put in alphabetical order by the last name of the author (or first-named author if more than one) or, if the author is not given, by the first main word of the title. Normally, you do not need to include works that you read, but did not actually use. A reference manager program for your computer can help to keep your references organized and in the proper format.

Find out which style sheet (MLA, APA, University of Chicago, etc.) your instructor prefers and conform to that format.

B. Preparing Final Source Citations

Whenever you borrow the words, facts, or ideas of others, you must acknowledge this within your text so that your reader will know where these things originated. Even if you do not directly quote your source (but only paraphrase or summarize it), you still need to cite it. Otherwise, you will be committing plagiarism (see chapter 1, "References," p. 26). Be sure to insert, parenthetically within your text, citations for all your sources. Again, the precise format you use will depend on which style sheet you are following.

Chapter Four

Writing Essay Exams

There are two types of essay exams: the short-answer essay exam and the complete essay one. For a short-answer essay, you must write from four to eight sentences; a single exam can contain many short-answer essays (for example, six to eight questions). For a complete essay exam, you are expected to write a timed essay approximately one to four pages in length; one exam will generally consist of no more than two such essays. Complete essays require a thesis statement and a development of your argument, as well as a conclusion. Both types of essay answers should be well organized and clearly presented. It is important to plan and outline what you will say *before you begin writing the exam*. Otherwise, your answer will be unclear and it will lack coherence and structure. This, in turn, will be reflected in your mark!

GETTING READY FOR THE EXAM

1. Take good lecture notes and review them regularly, and study the appropriate sections of the text and lab manual (if appropriate). Make condensed study notes that make sense to you.
2. Take note of which areas the instructor emphasized most strongly. Look for clues about what the instructor thinks is important.
3. Practice synthesizing what you know by creating summaries that help you recast the ideas of others in your own words and extract meaning

from notes and texts. Prepare outlines that reorganize the course material.

4. Ask questions of the course material that force you to examine the relations between issues. Focus on COMPONENTS (e.g., main parts of an issue, definitions of terms or theories), CHANGE (e.g., causal elements), and CONTEXT (how certain issues, theories, or results in the course compare with others). The most useful form of review for an essay examination forces you to analyze, integrate, and/or evaluate information.

5. Find out everything you can about the exam. Ask the professor and instructors about the format, length, style, and method of evaluation. Look at previous tests/exams to gain an understanding of the types of past questions. Try to use the practice tests in a timed setting at home without consulting any notes. In this way, determine your areas of strength and weakness. Compare your answers to the ones on the answer sheet to find out to what extent your answers were complete and well structured.

6. When reviewing the lecture material, ask yourself, "If I were the instructor, what would be the main four points that I would want students to understand?" and "Through what kinds of questions could the instructor test this material?"

WRITING SHORT-ANSWER ESSAY EXAMS

1. Preview the Exam: Spend 10% of the exam time reading the questions. Briefly look over all of them to gain an overall understanding of what is being asked and then determine which questions you feel most comfortable with.

2. Manage Your Time: Read the instructions carefully to find out how many questions you have to answer and see whether you have any choice. Subtract five minutes or so for initial planning, then look at the weighting of each question and divide the time accordingly, leaving some time at the end for rereading and editing each answer. Adhere to this time limit as you go along; otherwise you will run out of time!

3. Begin with the Easiest Questions: Start with the question that you feel most comfortable with, which may not necessarily be the first one. Continue through the exam, writing from the easiest to the hardest question. If you feel unable to attempt some questions, leave them to the last. Perhaps answering other questions will help trigger your responses to the more difficult ones.

4 Make an Outline for Each Answer: It is tempting, especially if you think you know the answer, to jump into writing without making a brief plan/outline beforehand. The lack of a plan, however, usually results in an unclear, confused answer that is difficult to read. Write down the key words, especially the verbs in the question (e.g., explain, compare, outline, discuss, review) and think about what, precisely, is being asked. *Misunderstanding the question is a common error!* Jot down key ideas related to the topic on rough paper or on the unlined pages of your answer book. Arrange these ideas into a brief outline. Now reread the question and make sure that your points are, in fact, relevant to the question.

5. Write Your Answer in Sentence Form: Be direct: state your key points in the first sentence and use the rest of the short answer to develop them. (Don't worry about a graceful introduction.) Try to support your answer frequently with examples from either lectures or the course textbook. Make clear transitions from sentence to sentence. Write legibly and on alternate lines. If you are a sloppy writer, then print. *Be certain that you are answering the original question.* Keep it clearly in front of you!

6. Leave Time to Review Each Answer: Leave about 10% of the time allotted to review your answers. Check especially for clarity of expression, content, and organization. Try to get rid of confusing sentences and increase the logical connections between ideas. Revision that makes answers easier to read is always worth the effort.

7. What If I Run Out of Time? Write down at least an outline in point form; it may be worth part marks, which is better than nothing at all!

WRITING COMPLETE ESSAY EXAMS

Complete essays may be approximately a full page or more in length. The above points apply; in addition, here are a few extra tips:

- Write an introduction containing your main hypothesis or point(s). In addition, briefly summarize the main supporting arguments.
- In each subsequent paragraph, develop these ideas further, using examples to support each argument.
- Be aware of the organization of your essay. You may want to arrange your points in increasing order of importance, a structure that leads naturally to a good conclusion.
- Ensure that your transitions are clear. This is particularly important in longer essay exams. Each paragraph should be logically linked to the previous one. One way to improve paragraph transitions is to foreshadow in a paragraph's final sentence what is about to be written in the next paragraph.
- Use key "landmark" or "signal" phrases to make your essay more "reader friendly." For example:
  - "The main purpose of gel electrophoresis is to separate nucleic acids or proteins by size."
  - "There are three main uses of electrophoresis ..."
  - "A common example is ..."
  - "Johnson and colleagues believe that ...; however, Hamilton contradicts this by ..."
  - "The conclusion from this experiment was ..."
- Indicate your conclusion in the last paragraph, ensuring that it is strongly and concisely stated.
- Avoid adverbs such as "clearly," "obviously," "undoubtedly," "unequivocally." What seems obvious today may not be so obvious tomorrow. Or, alternatively, what is clear to one scientist may not be so clear to another, or to the person marking the exam.
- Avoid vague quantifiers such as "more" and "greatly." Instead, state things as precisely as possible. For example, avoid: "The rate of transcription was more in the wild type than in the mutant." A better, more

concrete way of stating this is: "The rate of transcription was increased five-fold in the wild type compared to that in the mutant."

- Avoid wordiness and overly complex language and terminology. Stick to simple, yet effective, words. For example, employ "use" instead of "utilize." Do not be afraid to signal your points obviously, i.e., "The five main points are ..."

WHAT IS THE QUESTION ASKING?

Here are a few "key" words found in essay exam questions. These words are called "directives" because they direct the approach a student should take in answering the question.

Compare and contrast: Examine similarities between two key issues and contrast their differences.
Criticize: Judge and discuss the merits and faults of (i.e., "critique").
Defend (argue for): Find and marshal evidence to support the theory. Try to arrange your points from weakest to strongest.
Define: Explain or identify the nature or essential qualities of something. Include examples and diagrams if appropriate.
Design an experiment: Think carefully about the question and decide on the best experiment(s) that will lead to an answer. If several experiments are needed, make it clear how each experiment contributes to the overall result. Diagrams may be extremely useful. This type of question tests your ability to think critically and to apply any tools you have learned. Be sure to include controls in your experiments!
Enumerate: List various events, ideas, laws, etc.
Evaluate: Appraise the worth of a theory, procedure, etc. and justify your conclusion.
Explain: Make the meaning of something clear, plain, intelligible, and/or understandable.
Interpret: Give the meaning of something by paraphrase or by translation. Often, the concept you are supposed to interpret may be new to you. Use examples to support the interpretation that you make.

Illustrate: Use specific examples or analogies to clarify or explain.

Outline: Do a general sketch, account, or report, indicating only the main features of a book, subject, or project.

Review: Survey or summarize a topic, occurrence, or idea, generally but critically. Support with examples.

Summarize: State in concise form, omitting examples, analogies, and details.

Chapter Five

Let's Talk Sentences: Grammar Basics and the ABC's of Scientific Style

Until now, we have been dealing with the report, essay, poster, or proposal as a whole, trying to provide a *macro*-sense of how to approach a writing project and what your reader (or professor) expects. At this point, we want to shift gears somewhat and focus on the *micro*-level, examining the "nuts and bolts" of writing – the actual make-up of the phrase, the clause, the sentence, the paragraph. We will look at some basic grammatical concepts and frequent errors in grammar and punctuation, as well as consider some general principles of scientific style – ways in which you can most clearly, pleasingly, and persuasively express your ideas in a paper, especially a scientific one.

EVERYTHING YOU'VE ALWAYS WANTED TO KNOW ABOUT GRAMMAR BUT WERE AFRAID TO ASK

The study of grammar deals with the way in which words are arranged into meaningful patterns; it thus helps to explain how language works. It describes a language's usage, as well as its system of word classes – their inflections, functions, and relations in a sentence. Some people even consider grammar a science all to itself.

But don't worry. You do not need to be a rocket grammarian to write well; many writers (with no formal grammar training) manage to compose correct (and even eloquent) sentences and paragraphs – even entire essays and articles! Nevertheless, some basic familiarity with elementary grammatical concepts and principles *may* help you in your writing, especially if you are one of those who often encounters comments like "grammar?" or "agreement?" or "dangling" scrawled messily in the margins of your writing assignments.

Sentences, Clauses, and Phrases: The "Building Blocks"

A grammatically meaningful group of words, expressing a complete thought and known as a *sentence*, must contain both a *subject* and a *predicate*.

1. The *subject* of any sentence *identifies* or *names* what the sentence is about and is, in most cases, identifiable as a **noun** or **pronoun** (or, more rarely, a **noun clause** or **noun phrase**). Only very rarely is the subject implied or understood; e.g., "Handle with care" (implied subject: "you").
2. The *predicate* of any sentence *asserts* or *asks* something about the subject, or it tells the subject to do something (in other words, it specifies action about it). The essential part of any predicate is a **verb**; every predicate *must* contain a verb.
    Many patients undergo ligament reconstructive surgery every year.
    (subject: Many *patients*; predicate: undergo ligament reconstructive surgery every year; verb: undergo)
    These plants have developed a two-part photorespiration system.
    (subject: These *plants*; predicate: have developed a two-part photorespiration system; verb: have developed)

A *clause* is a group of words that, like a sentence, must contain a *subject* and a *predicate*. Although all sentences contain at least one clause, not all clauses are sentences. For example, some clauses, instead of making an independent statement, serve only as a subordinate part of the main sentence. Such clauses, called *dependent* (or *subordinate*) *clauses*, contain a subject and predicate but cannot stand alone as a sentence. *Independent clauses*, by contrast, can and do stand alone as complete sentences.

"Even though Darwin hypothesized the theory of sexual selection, he had no explanation for the initial evolution and maintenance of sexually selected traits."

(*dependent/subordinate clause*: "Even though Darwin hypothesized the theory of sexual selection,")

(*independent clause*: "he had no explanation for the initial evolution and maintenance of sexually selected traits.")

Independent and *dependent clauses* – and the various combinations thereof – are the component elements of the four different kinds of sentences. A **simple sentence** contains only one clause (an independent one); a **compound sentence** consists of two or more independent clauses, linked by coordinating conjunctions or punctuation marks; a **complex sentence** has one independent clause and one or more subordinate clauses; and a **compound-complex** sentence consists of two or more independent clauses, as well as one or more subordinate clauses.

Finally, a *phrase* is a group of words that functions like a single word or like a grammatical unit, even though it does not contain both a *subject* and a *predicate*.

"During contraction, each actin-myosin unit, called a sarcomere, pulls on the cell membrane, causing the cell to compress into a spherical shape."

(*phrase*: "causing the cell to compress into a spherical shape.")

The basic idea, when writing, is to compose your thoughts in complete sentences (which contain both a subject and a predicate) and to make sure that you properly "connect" or mark the breaks between/among the various clauses within your sentences. Writers commonly commit two errors in constructing sentences:

Fragments/Incomplete Sentences

A group of words that lacks a subject and predicate and does not express a complete thought, but is written as a sentence (i.e., begins with a capital letter and ends with a period), is called a "fragment." A "fragment" is typically either a **dependent clause** or a **phrase lacking a verb**.

(1) Before the absorbance of any solution was read. (dependent clause – fragment)

CORRECTED: Before the absorbance of any solution was read, the spectrophotometer was calibrated with a cuvette containing the blank solution.

(2) This experiment dealt with three mutant types of *Drosophila melanogaster*. For example, serrated wing mutants, curly wing mutants, and stubble hair mutants. (phrase, lacking verb – fragment)

CORRECTED: This experiment dealt with three mutant types of *Drosophila melanogaster* – serrated wing mutants, curly wing mutants, and stubble hair mutants.

Run-on Sentences/Comma Splices

A run-on sentence (or "fused sentence") occurs when two sentences (or independent clauses) "run together," with no adequate punctuation or coordinating conjunction to mark the break between them (such as a period, semi-colon, colon, or comma *and* a coordinating conjunction – i.e., "and," "but," "yet," "for," "or," "nor," "so").

(1) In some cases, the rats were removed from the maze after they ate all the fruit loops in other cases, they were removed after 10 minutes, no matter what they had eaten.

(The first clause ends after "fruit loops"; the second one begins with "in other cases.")

CORRECTED: In some cases, the rats were removed from the maze after they ate all the fruit loops; in other cases, they were removed after 10 minutes, no matter what they had eaten. (semi-colon inserted between the two clauses)

In many run-ons, the clauses are joined (or spliced) mistakenly *only* with a comma.

(2) The first test session was a PI test in which the rats received three trials on Maze A, the second test session – conducted to show release from PI – involved two runs on Maze B, followed by a run on Maze A.

(A comma is insufficient to join these two independent clauses.)

CORRECTED: The first test session was a PI test in which the rats received three trials on Maze A. The second test session – conducted to show release from PI – involved two runs on Maze B, followed by a run on Maze A. (period used instead of comma to separate the two clauses)

Common Errors in Grammar/Punctuation and How to Fix Them

Here is a list of some other common errors in grammar and punctuation. This is not a complete list, but it gives you some possible problems to watch for as you proofread your own work.

1. Faulty subject-verb agreement

A verb must agree with its subject in number (singular or plural) and in person (first, second, or third). Always check to make sure that your verb agrees with your subject.

Two *replicates* of each treatment was calculated. (✗)
("replicates" is the subject [plural]; verb form should be plural: "were")

All priapulids lives in unusual, harsh, or marginal environments. (✗)
("priapulids" is the subject [plural]; verb form should be plural: "live")

In many species, the *male*, as well as the female, *care* for the offspring. (✗)
("male" is the subject [singular]; verb form should be singular: "cares")

The number of specimens tested by the lab *exceed* 3 000. (✗)
("the number of ..." as subject takes a singular verb form, i.e., "exceeds")

Our *data indicates* that the level of production of the new gene products is not sufficient to mask the expression of other genes for the same trait. (✗)
(nouns such as "data," "criteria," "media," "trivia," "strata," and "phenomena" are plural; they always require a plural verb – i.e., "indicate")

Either the molecular weights of the protein or the *distance* it travels *are* difficult to determine. (✗)
(for compound subjects whose parts are joined by "or," "nor," "but," "either ... or," "neither ... nor," or "not only ... but also," the verb agrees in number and person with the part of the subject closest to the verb; in this case, the subject "distance" [closest to the verb] is singular, so the verb should be "is")

This solution has many immediate promises, including the elimination of immunological *reactions*, which currently *hampers* the use of allografts as a possible solution. (✗)

(in a relative clause, the verb must agree with the antecedent of the relative pronoun [i.e., the antecedent of "which" is "reactions," so the verb must also be plural – "hamper"])

2. Incorrect verb tenses

Tenses are the forms of a verb that show when the action or condition expressed by the verb takes place or exists. In scientific writing, the most commonly used verb tenses are the present, past, present perfect, and future.

Use the **present** tense to summarize the general scope or direction of your paper or to indicate either actions or conditions occurring at the time of speaking or writing or those considered to be general truths or scientific facts.

This experiment *has investigated* [CORRECT: *investigates*] how the rate of the electron transport chain is dependent on the amount of mitochondrial extract isolated.

Figure 4 *will compare* [CORRECT: *compares*] stimulus frequency and muscle response.

Conjugation *is involving* [CORRECT: *involves*] the transfer of genetic material between cells.

Use the **past** tense to indicate actions or conditions that occurred at a specific time and do not extend into the present.

The angular separation of the light *has been made* [CORRECT: *was made*] at several different frequencies. (From "Methods & Materials")

Sometime before 1000 CE, the Chinese *had discovered and developed* [CORRECT: *discovered and developed*] pyrotechnic devices, or fireworks.

Use the **present perfect** tense to indicate actions or conditions begun in the past and either completed at some unspecified time in the past or continuing into the present.

> Although many exact and approximate results *were* [CORRECT: *have been*] obtained from the theory of quarks and gluons (QCD), the mathematics *were* [CORRECT: *are*] quite difficult.

> The ability to mutate specfic bases by site-directed mutagenesis (SDM) *was* [CORRECT: *has been*] a powerful tool to characterize genes since its introduction in the 1980s.

> Advances over recent decades *showed* [CORRECT: *have shown*] that Newton's laws are inadequate at the atomic scale and at velocities comparable to the speed of light.

3. Unnecessary shifts in verb tense, mood, and voice

Be careful not to change verb tenses, mood (indicative vs. imperative), or voice (passive vs. active) unnecessarily.

> Approximately 100 seeds *were* transferred [PAST TENSE] to the tube, and a pipette *is* used [PRESENT TENSE] to add a small amount of sterile solution. (✗)
> REVISED: Approximately 100 seeds *were* transferred to the tube, and a pipette *was* used to add a small amount of sterile solution.

> Hook [IMPERATIVE MOOD] an oxygen electrode up to a polarographic computer program; you *should* then *calibrate* [INDICATIVE MOOD] the electrode to detect and record any changes in oxygen concentrations. (✗)
> REVISED: *Hook* an oxygen electrode up to a polarographic computer program; *calibrate* the electrode; and *record* any changes in oxygen concentrations.

> Once specialists fully *understood* [ACTIVE VOICE] the structure of hemoglobin, the first hemoglobin solution *was made* [PASSIVE VOICE] in 1937. (✗)
> REVISED: Once the structure of hemoglobin *was* fully *understood*, the first hemoglobin solution *was made* in 1937. [See "Scientific Writing Style," "Conciseness," "Passive versus Active Voice," below.]

4. Faulty predication

When the subject of a sentence is not grammatically connected to what follows (the predicate), the result is **faulty predication**.

> The *reason* the equilibrium constant for this reaction is high is *because* H+ is a much stronger acid than HOAc. (✗)
>
> (The subject, "reason," needs a "that" clause to complete it; the conjunction "because" [meaning "for the reason that"] is redundant.)
>
> CORRECTED: The reason the equilibrium constant for this reaction is high is that H+ is a much stronger acid than HOAc.
>
> *or*
>
> The equilibrium constant for this reaction is high because H+ is a much stronger acid than HOAc.
>
> (The second correction is preferable because it is less wordy.)
>
> The *photoelectric effect* is *when* light bombards the surface of a metal and electrons are ejected. (✗)
>
> CORRECTED: The photoelectric effect occurs when light bombards the surface of a metal and electrons are ejected.

5. Faulty pronoun agreement

A pronoun should agree in number, person, and gender with the noun to which it refers (its antecedent).

> We investigated a frog quadricep *muscle* [ANTECEDENT: singular] and their [CORRECT: *its*] response to electrical stimulation.
>
> The grey *squirrel* [ANTECEDENT: singular], *Scirius carolinesis*, is a small, common forest-dwelling mammal. *Their* [CORRECT: *Its*] food source is almost exclusively nuts.
>
> Neither the twitch maximum nor the sub-tetanus *contractions* [ANTECEDENT: plural] demonstrated *its* [CORRECT: *their*] accurate measurement.
>
> A *scientist* [ANTECEDENT: singular – masculine or feminine] must refrain from expressing *their* [CORRECT: *his or her*] emotional reactions.
>
> *or*

Scientists [ANTECEDENT: plural] must refrain from expressing *their* emotional reactions.

6. *Dangling modifiers*

Be sure that a modifying phrase or clause (participial or infinitive) has something to modify. A dangling modifier is one that does not sensibly modify anything in the sentence. Check all words ending in "ing" and see whether they need a subject. If they do, is that subject correctly and unambiguously identified? Because of the prevalence of the passive voice in scientific writing, dangling modifiers are common and sometimes very difficult to revise!

Studying [DANGLING PARTICIPLE] these areas, the conclusions were obvious. (WRONG: This implies that the conclusions were doing the studying!) REVISED: When the scientists studied these areas, the conclusions were obvious.

Once the seeds are settled, the water is aspirated, leaving the seeds at the bottom. (WRONG: Who or what is leaving the seeds at the bottom? The seeds? The water?) REVISED: Once the seeds are settled, the water is aspirated so that the seeds remain at the bottom.

To verify [DANGLING INFINITIVE] the theory, open supracondylar humeral fractures were treated by open reduction and internal fixation. (WRONG: This implies that the fractures themselves were verifying the theory!) REVISED: To verify the theory, researchers treated open supracondylar humeral fractures by open reduction and internal fixation.

7. *Misplaced modifiers*

Place modifiers where they will clearly modify the words intended. Avoid "squinting modifiers" (ones that may refer to either the preceding or the following word).

In 1975, Des Collins of the Royal Ontario Museum mounted an expedition in and around both quarries to collect fossils from the debris slopes. (*in and around both quarries* properly modifies "debris slopes.")
REVISED: In 1975, Des Collins of the Royal Ontario Museum mounted an expedition to collect fossils from the debris slopes in and around both quarries.

The quantization of electron spin means that there are two possible orientations only of an electron in a magnetic field, one associated with a spin quantum number. (*Only* is a squinting modifier. Does *only* refer to "two possible orientations" or to "of an electron"?)
REVISED: The quantization of electron spin means that there are only two possible orientations of an electron in a magnetic field, one associated with a spin quantum number.

8. Misuse of punctuation marks: commas, semicolons, and colons
Use a **comma** after each item in a series of three or more. Its presence ensures clarity when you are listing a series of nouns or actions.

Parkinson's disease is characterized by muscular rigidity, resting tremor, the slowing of movement and the blunt affect of its patients. (✗)
REVISED: Parkinson's disease is characterized by muscular rigidity, resting tremor, the slowing of movement, and the blunt affect of its patients.

Use a **comma** to separate coordinate adjectives before nouns (even if there are only two); if the adjectives are *not* coordinate, do *not* use a comma. To check whether they are coordinate or not, insert "and" between them or reverse their order. If the phrase still makes sense, the adjectives are coordinate. If it does not, they are not coordinate and no comma is necessary between the adjectives.

Scientists found evidence of a *specialized, breathing* device called a hepatic piston. (✗: non-coordinate adjectives – no comma needed)
REVISED: Scientists found evidence of a specialized breathing device called a hepatic piston.

Use a **comma** before and after "non-essential" elements in a sentence to set them apart from the rest of the sentence. "Non-essential elements" include all clauses, phrases, and appositives that are not essential to defining the meaning of the words or phrases they refer to. Do not use a comma before and after "essential" elements.

Several sources of error which may or may not have altered the results will be eliminated in the next trial.

(✗: non-essential clause "which may have altered the results" should be set off by commas.)

REVISED: Several sources of error, which may or may not have altered the results, will be eliminated in the next trial.

The wave motion, we have described so far, is that of traveling waves.

(✗: the clause "we have described so far" is essential to define "wave motion" and should not be set off by commas.)

REVISED: The wave motion we have described so far is that of traveling waves.

Use a **comma** to separate two independent clauses joined by a coordinating conjunction (and, but, or, nor, for, yet, so).

The group obtained proof of the substance's chemical composition but too little of the white crystalline powder was present to use in the experiment.

(✗: sentence contains two independent clauses joined by coordinating conjunction "but")

REVISED: The group obtained proof of the substance's chemical composition, but too little of the white crystalline powder was present to use in the experiment.

Aluminum is the third most abundant element in the earth's crust, and has many important uses in our economy.

(✗: sentence contains only one independent clause [but has a compound predicate], so no comma is necessary)

REVISED: Aluminum is the third most abundant element in the earth's crust and has many important uses in our economy.

Do not use a comma to separate subjects and verbs or to separate verbs and objects.

Developing effective preventive strategies and therapies, depends on understanding the critical genetic and corresponding phenotypic changes that
occur in the course of breast carcinogenesis.

(✗: comma is separating the subject "Developing ..." and the verb "depends")

REVISED: Developing effective preventive strategies and therapies depends
on understanding the critical genetic and corresponding phenotypic
changes that occur in the course of breast carcinogenesis.

Use a **semi-colon**, not a comma, to join independent clauses without a coordinating conjunction.

Otherwise, hydrolysis may occur near the top of the solution, where the substrate was added, thus, the absorbance read will not reflect the real rate of
activity. (✗)

REVISED: Otherwise, hydrolysis may occur only near the top of the solution,
where the substrate was added; thus, the absorbance read will not reflect the
real rate of activity.

Use a **colon**, not a semi-colon, to introduce a sentence that is an explanation
or example of what precedes it. (Avoid putting colons after expressions
such as "such as," "especially," and "including.")

C3 plants function in the following manner; carbon dioxide diffuses from
the air, where it is abundant, to the leaf, where it is scarce.

(✗: second clause, beginning with "carbon dioxide," is an explanation of first
clause, beginning with "C3 plants")

REVISED: C3 plants function in the following manner: carbon dioxide diffuses
from the air, where it is abundant, to the leaf, where it is scarce.

SCIENTIFIC WRITING STYLE – THEORY AND PRACTICE

In terms of style, we want to stress three main principles: (1) clarity, (2) con-
ciseness, and (3) forcefulness. We also wish to propose some practical ways
in which you might achieve these principles in your writing.

Clarity

Choose clear words.

Use a good dictionary when you write. Look up the exact meanings of any words you are unsure of. *Webster's New International Dictionary* gives good up-to-date definitions of technical terms; you might also consult a specialized science dictionary in your field. Consult a thesaurus with care. A thesaurus can be useful when you want to avoid repeating words or when you are searching for a precise word. But make sure you get the correct **connotation** (the particular overtones, emotional colourings, or associations) of the word in question.

Avoid jargon and use plain English whenever possible.

All disciplines have their specialized terminology; this vocabulary is often essential, but writers too often use jargon unnecessarily. Use jargon only when it helps to explain something more precisely or efficiently. Never use it to sound more knowledgeable.

Plain words are almost always clearer than fancy ones. Beware of words with a lot of prefixes (pre, post, anti, pro, sub, etc.) or suffixes (ate, ize, tion). Sometimes these words can make your writing more precise; too often they only make it dense and hard to understand. Use plain substitutes for words that are unnecessarily complex (as in the following table).

Unnecessarily complex words	Plain substitutes
determinant	cause
cognizant	aware
obviate	prevent
requisite	needed
utilization	use
terminate	end
prior to	before
finalize	complete

Use pronouns clearly.

Ensure that any pronouns you use refer clearly to their antecedents – the nouns or clauses they replace. Pronouns like "it," "they," "them," "this," "that," or "which," particularly at the beginning of a sentence, should refer unambiguously to a preceding noun or clause. If they do not, you will confuse your reader and your meaning will be lost. They should also agree in number (singular or plural) with their antecedents.

What, exactly, do "they," "this," and "which" refer to, or mean, in these sentences? Note the ambiguity until the sentence is properly revised.

1. These solutions were prepared in regular test tubes and later transferred to spectrophotometric cuvettes for reading. *They* were also sterilized. ("solutions," "regular test tubes," or "cuvettes"?)
2. The Kunkel method uses uracil-containing DNA that is rapidly degraded in wild type *Escherichia coli*. *This* prevents the inheritance of a parental wild type vector in the progeny. ("the DNA," "*Escherichia coli*," or "process of degradation"?)

REVISED:

1. These solutions, which were prepared in regular test tubes and later transferred to spectrophotometric cuvettes for reading, were also sterilized.
2. The Kunkel method uses uracil-containing DNA that is rapidly degraded in wild type *Escherichia coli* and, thus, prevents the inheritance of a parental wild type vector in the progeny.

Clarify comparatives.

When you use comparatives such as "more," "less," "greater," "better," etc., always ask yourself "than what?" and make the answer clear to your reader. Ensure that your comparisons are complete and unambiguous.

This experiment is *less difficult*, and we will, therefore, not need the full thirty days. (AMBIGUOUS)

REVISED: This experiment is less difficult than the previous one, and we will, therefore, not need the full thirty days.

Trout eat *more algae than minnows.* (AMBIGUOUS)
REVISED: Trout eat more algae than minnows do.

Also, the second element of any comparison should always be equivalent to the first.

Jacobovici's results are *erroneous findings,* as is his *hypothesis.* (NON-EQUIVA-LENT)
REVISED: Jacobovici's results are erroneous, as is his hypothesis.

The *laboratory performance* of Fajan's method is more effective than *Lazebnik or Mohr.* (NON-EQUIVALENT)
REVISED: The laboratory performance of Fajan's method is more effective than that of either Lazebnik or Mohr.

Limit the number of negatives.

Do not use too many negative terms in the same sentence. Two negatives can be used to refer to alternatives, for example, "neither A nor B is visible." But in grammar, as in math, two negatives usually make a positive. In addition, overused negatives may make sentences difficult to understand.

For example, try to decipher the meaning of the following:
Only in Pearlman's experiments were there never sufficient data to disprove Bender's hypothesis.

Use correct prepositions.

Prepositions (e.g., "at," "in," "for," "on," "by," "of," "from," "to," "than," "through") indicate relationships between nouns. Some verbs take one preposition and one preposition only. (We can say only "I am tired of you," not "I am tired from you.") Check in a dictionary if you are unsure of which preposition to use with a verb. (One good dictionary that has a section on which preposition follows a particular verb is the *Collins Cobuild English Guide: 1 (Prepositions).* Another dictionary devoted entirely to prepositions is Frederick Wood's *English Prepositional Idioms.*)

Do not use "different than" if you can use "different from." If you use the word "following" as a preposition, it often sounds like a dangling participle. It is usually better to substitute "after."

> The rats were anesthetized following the period of observation. (WEAK)
> REVISED: The rats were anesthetized after the period of observation.

Do not overuse nouns as modifiers.
Avoid what is known as "noun strings" or "noun-adjective clusters."

How would you simplify or clarify the following?:

AWKWARD: a steroid-induced GAB channel burst duration prolongation
REVISED: a steroid-induced prolongation of the burst duration of GAB-activated channels

AWKWARD: electronic microscope utilization manual
REVISED: manual for using electronic microscopes

AWKWARD: air pollution investigation committee
REVISED: committee to investigate air pollution

Conciseness

Use adverbs and adjectives sparingly.
One well-chosen modifier is always better than a long series of synonyms. Remove unnecessary adjectives and adverbs, especially vague qualifiers such as very, quite, rather, fairly, relatively, comparatively, several, much. "Very" reduces the impact of the term you are trying to strengthen.

Notice the unnecessary qualifiers in the following sentences and the more concise revisions:

> The enzyme can also be made to select *fairly efficiently* one isomer, or molecular form of a compound, from a mixture of *various* forms to produce one *single* isomer of the product.

REVISED: The enzyme can also be made to select one isomer, or molecular form of a compound, from a mixture of forms to produce one isomer of the product.

The cells can *often* be immobilized on a supporting structure for *relatively* continuous processing.

REVISED: The cells can be immobilized on a supporting structure for continuous processing.

Avoid chains of relative clauses.

Sentences full of clauses beginning with "which," "that," or "who" are usually more wordy than necessary. Try reducing some of the clauses to phrases. For example:

Electrophoresis is an efficient technique that is used to separate and analyze the mixtures of proteins which are found in human blood and other biological materials.

REVISED: Electrophoresis is an efficient technique for separating and analyzing the mixtures of proteins found in human blood and other biological materials.

One of the main problems that is faced at this point in time is that there is a reduction in the ozone layer.

REVISED: One main problem now is the reduced ozone layer.

Try reducing independent clauses to phrases or words.

Independent clauses can also often be reduced by what we call "subordination."

The abstract was written clearly and concisely [independent clause], and it was widely read.

REVISED: Written clearly and concisely [phrase], the abstract was widely read.

Eliminate clichés.

Get rid of clichés, colloquialisms, and wordy introductory phrases such as "in this connection we may say that ..." Here are some examples:

WORDY:	BETTER:
due to the fact that	because, since
at this point in time	now
consensus of opinion	consensus
in all likelihood	likely
has the capability of	can, is able to
in order to	to
in the near future	soon
in the eventuality that	if
fewer in number	fewer
with reference to	about
it is worth noting that	OMIT
it is interesting to note that	OMIT
as mentioned before	OMIT

Forcefulness

Attempt to develop a forceful, vigorous style and learn how to emphasize your key ideas. If you do so, you will communicate more effectively with your readers because your ideas will be more persuasive. Here are some ways to accomplish this goal.

Focus on the real subject.
The "real" subject of a sentence should also be its *grammatical* subject, and it should appear prominently. If you bury the "real" subject in a weak phrase, you weaken the meaning and emphasis in your sentence. Also, when possible, eliminate expletives such as "it is," "there are," and "there is." These words only clutter your text and de-emphasize your main ideas.

Note the real subject in the following examples:

(1) The use of this method would eliminate calibration inconsistencies.
REVISED: *This method* would eliminate calibration inconsistencies.

(2) There has been no study yet that reports the outcome of open-only distal humeral fractures that have been treated by open reduction.

REVISED: *No study* has yet reported the outcome of open-only distal humeral fractures that have been treated by open reduction.

(3) It is the skeletal muscle that is excited when acetylcholine is released from a somatic motor neuron.

REVISED: The *skeletal muscle* is excited when acetylcholine is released from a somatic motor neuron.

(4) The presence of a vertical lesion was observed.

REVISED: A *vertical lesion* was observed.

Focus on the "real" verb.

A "real" verb should also be evident in the sentence. Try not to "nominalize" your verb (i.e., turn it into a noun). For example, do not use "installation" instead of the verb "install." The more nouns in your text, the harder it is to read.

> Each *preparation* of the solution was *done* twice. (noun)
> REVISED: Each solution was *prepared* twice.

> An *investigation* of all possible alternatives was *undertaken*. (noun)
> REVISED: All possible alternatives were *investigated*.

> *Consideration* should be *given* to retesting Asinovsky's findings. (noun)
> REVISED: We should *consider* retesting Asinovsky's findings.

Use the passive or active voice as appropriate.

"Voice" is a feature of transitive verbs (verbs that take a direct object) that tells whether the subject is acting (*he questions us*) or is being acted upon (*he is questioned [by us]*). When the subject is acting, the verb is in the *active voice*; when the subject is being acted upon, the verb is in the *passive voice*.

Although in many disciplines the active voice is preferred (since it is more direct, forceful, and personal), much scientific writing uses the passive effectively to highlight the object or phenomenon being studied rather than the person or persons doing the studying (for example, in the Introduction, Materials and Methods, Results, and Abstract sections).

e.g., The experiment *was performed* to investigate how various processes affect the respiratory chain.

When, however, a sentence is awkward or unnatural because of the passive voice, change the verb from the passive to the active form.

Differences between the hexokinases of C. *utilis* and S. *cerevisiae*, mainly concerning physiological regulation for both growth source and effectors, *have been pointed out* by previous works. (PASSIVE VOICE)

REVISED: Previous works *have pointed out* some differences between the hexokinases of C. *utilis* and S. *cerevisiae*, mainly concerning physiological regulation for both growth source and effectors. (ACTIVE VOICE)

Use the passive voice when

- the agent (or doer of the act) is indefinite or unknown;
- the agent is less important than the act itself (which is common in scientific writing);
- you want to emphasize either the agent or the act by putting it at the beginning or the end of the sentence.

There are, however, occasions when the active voice is preferable. See "Use personal subjects ..." below.

Follow a grammatical subject as soon as possible with its verb.
Readers expect a grammatical subject to be followed immediately by the verb. Anything lengthy that intervenes between subject and verb is read as an interruption, and therefore as something of lesser importance. Long sentences are not necessarily difficult to read; it is the structure of the sentence that counts.

Use parallelism for balance and emphasis.
Parallelism requires that similar ideas be presented in similar form and that elements that are similar in function appear in similar grammatical form. Stylistically, parallelism creates balance and emphasis. Always be watching for "faulty parallelism."

While falling, the cat extends its forelegs and the hind portion is rotated. (switch from active to passive voice)

REVISED: While falling, the cat extends its forelegs and rotates its hind portion.

The new refrigerant not only decreases energy costs, but also spoilage losses. (not only ... but also ...)

REVISED: The new refrigerant decreases not only energy costs but also spoilage losses.

The titration process is producing errors and creates inconsistencies. (switch from present progressive to simple present)

REVISED: The titration process produces errors and creates inconsistencies.

Use personal subjects and concrete details to bring facts to life.

The more concrete and direct your language, the more forcefully (and persuasively) you will communicate with your reader. In particular, when you are writing non-scientific reports and/or essays, use personal subjects and active verbs whenever possible.

The materialistic *implications* of Darwin's theory led to a long delay before it was published. (WEAK: "Implications" is an abstract noun; "was published" is in the passive voice.)

REVISED: *Darwin* delayed publication of his theory for a long time because of its materialistic implications. ("Darwin" is a concrete noun; "delayed publication" is an active construction.)

Emphasize important ideas.

Place key words in strategic positions, subordinate minor ideas, and vary sentence length and sentence structure. Save your key words/ideas for the last.

An electric shock of minimum voltage is applied [independent clause], and the disturbance of the sarcolemma causes a twitch contraction of the muscle. (WEAK: "the disturbance of the sarcolemma causes a twitch contraction of the muscle" is not emphasized since it is only one of two independent clauses.)

REVISED: When an electric shock of minimum voltage is applied [dependent clause], *the disturbance of the sarcolemma causes a twitch contraction of the muscle*. ("the disturbance of the sarcolamma causes a twitch contraction of the muscle" is now emphasized because it is the main clause of the sentence; the other clause is now a dependent one.)

Appendix 1:
Resources on Writing

Here is a very short, selective list of print and electronic resources. There are, quite literally, thousands of books and web sites on these topics. Most of the books on this list are probably available in the various libraries at your college and university, or at the university/college bookstore.

In addition, we urge you to make use of the many resources available to you at your university or college. In particular, an extremely valuable, free resource at most colleges and universities is the "writing center," "writing lab," or "writing workshop" (they are called by different names), where you can receive individual tutoring and/or group teaching to help you develop your writing skills. Writing centers accept students from all levels and backgrounds and on any type of writing assignments (except, usually, take-home examinations). Experienced writing tutors will work with you on a one-on-one basis, dealing with your individual writing assignments, no matter at what stage of the writing process you are (brainstorming, researching, outlining, writing, revising, etc.) A writing tutor can assess your particular needs and provide you with invaluable suggestions and strategies, tailor-made to your own purposes. *Check it out!*

HANDBOOKS ON SCIENTIFIC/TECHNICAL WRITING

Barass, R. (1978). *Scientists Must Write: A Guide to Better Writing for Scientists, Engineers, and Students.* Chapman & Hall, London; Wiley, New York. Provides assistance on lab reports, research reports, proposals, and oral presentations.

Bazerman, C. (1988). *Shaping Written Knowledge: The Genre and Activity of the Experimental Article in Science*. University of Wisconsin, Madison.

Blake, G., and Bly, R.W. (1993). *The Elements of Technical Writing*. Macmillan Publishing Co., New York. Readable and concise.

Booth, V. (1993). *Communicating in Science: Writing a Scientific Paper and Speaking at Scientific Meetings*. Cambridge University Press, Cambridge.

Brusaw, C.T., Alred, G.J., and Oliu, W.E. (1976). *Handbook of Technical Writing*. St. Martin's Press, New York.

Higham, N. (1993). *Handbook of Writing for the Mathematical Sciences*. Society for Industrial and Applied Mathematics, Philadelphia. Clear and practical advice on how to discuss numerical concepts in words.

Huth, E.J. (1999). *Writing and Publishing in Medicine*. Williams and Wilkins, Baltimore. An authoritative guide to writing in this specialized field.

Markel, M.H. (1994). *Writing in the Technical Fields: A Step-by-Step Guide for Engineers, Scientists, and Technicians*. IEEE Press, Piscataway, NJ; Institute of Electrical and Electronics Engineers, New York. Practical and detailed advice, particularly regarding difficulties in planning, drafting, and revising specific kinds of documents.

Northey, M., and Timney, B. (1995). *Making Sense in Psychology & the Life Sciences: A Student's Guide to Research, Writing, & Style (APA Format)*. Oxford University Press, Don Mills, Ontario.

O'Connor, M. (1991). *Writing Successfully in Science*. Chapman & Hall, London.

Pechenik, J. (1987). *A Short Guide to Writing about Biology*. Little Brown, Boston & Toronto.

STYLE MANUALS

Baker, S. (1986). *The Practical Stylist*. Harper & Row, New York.

Council of Biology Editors, Style Manual Committee. (1994). *Scientific Style and Format: The CBE Manual for Authors, Editors, and Publishers*. 6th ed. Cambridge University Press, Cambridge & New York. General style conventions; special scientific conventions; journal and book writing; the publishing process.

Hacker, D. (1997). *A Pocket Style Manual*. Bedford Books, Boston.

Kirkman, J. (1992). *Good Style: Writing for Science and Technology*. E & FN Spon, London.

Lane, J., and Lange, E. (1993). *Writing Clearly: An Editing Guide*. Heinle & Heinle, Boston. Designed for ESL learners.

Northey, M., and Procter, M. (1998). *Writer's Choice: A Portable Guide for Canadian Writers*. Prentice Hall Canada Inc., Scarborough, Ontario.

Rubens, P., ed. (1994). *Science and Technical Writing: A Manual of Style*. H. Holt, New York. Massive reference book; discusses audience analysis, numbers, symbols, etc.

Strunk, W.J., and White, E.B. (1979). *The Elements of Style*. 3rd ed. Macmillan Publishing Co., New York. Classic guide to clear and effective language.

Zinsser, William. (1998). *On Writing Well: the Classic Guide to Writing Non-fiction*. 6th ed. HarperCollins, New York.

DICTIONARIES, GRAMMAR, PUNCTUATION, WRITING, REFERENCING, AND FORMATTING

Collins Cobuild English Dictionary. (1995). HarperCollins, London. Differs from traditional dictionaries in that it provides grammatical information and an illustrative sentence for each meaning. Dictionary can also be consulted on the web at <http://titania.cobuild.collins.co.uk>

Eyring, J., and Frodesen, J. (1993). *Grammar Dimensions Four: Form, Meaning, and Use*. Heinle & Heinle, Boston.

Frank, M. (1993). *Modern English: A Practical Reference Guide*. Regents/Prentice Hall, Englewood Cliffs, NJ. Thorough grammar reference book.

Gelfand, H. (1994). *Mastering APA Style: Student's Workbook & Training Guide*. American Psychological Association, Washington, DC.

Hacker, D. (1996). *A Canadian Writer's Reference*. Nelson Canada, Scarborough, Ontario. Guidebook on correct style and format for academic writing. Good sections on grammar, punctuation, referencing.

Li, X., and Crane, N.B. (1996). *Electronic Style: A Guide to Citing Electronic Information*. Info Today, Medford, NJ.

Norton, S., and Green, B. (1997). *The Bare Essentials Plus*. Harcourt Brace, Toronto. Good grammar handbook on the basics.

Partridge, E. (1963). *You Have a Point There: A Guide to Punctuation and Its Allies*. H. Hamilton, London. Gives both historical and current usage of punctuation marks.

Publication Manual of the American Psychological Association. (1983). Washington, DC. The official APA documentation manual. Good on tables and diagrams.

Shertzer, M. (1986). *The Elements of Grammar*. Macmillan Publishing Co., New York.

Troyka, L.Q. (1999). *Simon and Schuster Handbook for Writers*. Prentice-Hall Allyn & Bacon Canada, Scarborough, Ontario.

Weissberg, R., and Buker, S. (1990). *Writing Up Research: Experimental Research Report Writing for Students of English*. Prentice-Hall/Regents, Englewood Cliffs, NJ. Chapters on abstract, introduction, methods/materials, results, and discussion sections.

SOME WEB SITES TO EXPLORE

Canadian Medical Association Medical Writing Centre <http://www.cma.ca/publications/mwc/index.htm>

Canadian Science Writers' Association
  <http://www.interlog.com/~cswa/index/html>
Engineering Writing Centre/U of T
  <http://www.ecf.toronto.edu/~writing/>
Health Sciences Writing Centre/U of T
  <http://www.utoronto.ca/hswriting/>   Several excellent writing handouts
Inkspot, Resources for Technical/Scientific Writers
  <http://www.inkspot.com/ss/genres/tech.html>   Resources for writers of all levels
  and experience.
NASA (U.S. National Aeronautics and Space Administration)
  <http://stipo.larc.nasa.gov/sp7084/sp7084cont.html>   Detailed on-line handbook on
  grammar, sentence structure, punctuation, and capitalization.
National Association of Science Writers
  <http://nasw.org>
Purdue University, Indiana, Online Writing Lab
  <http://owl.english.purdue.edu/writers/prose.html>   High-quality handouts on ESL,
  parts of speech, punctuation, resumé writing, etc.
Society for Technical Communication
  <http://stc.org>
University of Toronto
  <http://www.utoronto.ca/writing/>   Dozens of useful handouts on essay writing,
  critical reading, plagiarism, etc.

Appendix 2:
The Classification of Organisms

Each organism has its own unique name that is part of a larger group. All organisms can be classified into groups, according to an established hierarchy or taxonomy, on the basis of physical, ecological, and molecular characteristics, such as sequence variability. The seven major levels of taxonomy are kingdom, phylum (pl. phyla) or division, class, order, family, genus (pl. genera), and species (pl. species). Below, we enumerate the different kingdoms to explain the organization of species; we also provide a few examples of taxonomic classifications. Finally, we list the Latin names of common species used in scientific investigations.

In the Linnaean system of classification, there are *five kingdoms*:
1. **Monera:** small unicellular prokaryotes, e.g., cyanobacteria, bacteria.
2. **Protista:** large unicellular eukaryotes, e.g., protozoans and algae.
3. **Fungi:** unicellular, multicellular, and filamentous eukaryotes, e.g., molds, mushrooms, yeasts, mildew, and smut.
4. **Plantae:** multicellular, sessile (immobile) eukaryotes, and photosynthetic, e.g., mosses, ferns, woody and non-woody plants, and flowering plants.
5. **Animalia:** multicellular (with own means of locomotion), e.g., sponges, worms, insects, fish, birds, and mammals.

Examples of taxonomic classifications:
1. Dog
   Animalia; Chordata; Mammalia; Canidae; Carnivora; Canis; familiaris (note: you would mention only *Canis familiaris* in your report to indicate this species)

2. *Escherichia coli* Bacteria
   Monera; Bacteria; Proteobacteria; Enterobacteriaceae and related endo-symbionts; Enterobacteriaceae; Escherichia; coli
3. Corn
   Plantae; Anthophyta; Monocotyledones; Commelinales; Poaceae; Zea; mays
4. Edible Mushroom
   Fungi; Basidiomycota; Hymenomycetes; Agaricales; Agaricaceae; Agari-cus; bisporus.

In scientific literature, a binomial nomenclature, including both the genus and species names, is required to name the organism. The first letter of the genus name is capitalized, but not that of the species name. Latin names should be used, and both genus and species names italicized or underlined. For example, "dog" would be designated as *Canis familiaris*, and "corn" as *Zea mays*. Genus and species names *must* be fully spelled out on first mention and when used at the beginning of a sentence. At all other times, the genus name may be abbreviated, e.g., *A. bisporus, Z. mays*.

A quick guide to the *Latin names* of commonly used species in biology:

Algae
Chlamydomonas reinhardtii

Bacteria
Bacillus subtilis

Bacteria
Escherichia coli

Bacteria
Pseudomonas syringae

Bacteria
Salmonella typhimurium

Cockroach
Periplaneta americana

Corn
Zea mays

Fish
Fugu rubripe

Frog
Xenopus laevis

Fruit fly
Drosophila melanogaster

Human
Homo sapien

Mouse
Mus musculus

Nematode
Caenorhabditis elegans

Protist
Paramecium caudatum

Rat (Norway)
Rattus norvegicus

Rice
Oryza sativa

Slime mold
Dictyostelium discoideum

Tobacco
Nicotiana tabacum

Tomato
Lycopersicon esculentum

Thale Cress
Arabidopsis thaliana (plant)

Yeast
Saccharomyces cerevisiae

Yeast (fission)
Schizosaccharomyces pombe

Zebra fish
Danio rerio

Appendix 3:
International Units

SI base units

Quantity	Name	Symbol
Length	meter	m
Mass	kilogram	kg
Time	second	s
Electric current	ampere	A
Thermodynamic temperature	kelvin	K (not °K)
Amount of substance	mole	mol
Luminous intensity	candela	cd

Prefixes used with SI units

Factor	Prefix	Symbol	Factor	Prefix	Symbol
10^1	deca	da	10^{-1}	deci	d
10^2	hecto	h	10^{-2}	centi	c
10^3	kilo	k	10^{-3}	milli	m
10^6	mega	M	10^{-6}	micro	µ
10^9	giga	G	10^{-9}	nano	n
10^{12}	tera	T	10^{-12}	pico	p
10^{15}	peta	P	10^{-15}	femto	f
10^{18}	exa	E	10^{-18}	atto	a
10^{21}	zetta	Z	10^{-21}	zepto	z
10^{24}	yotta	Y	10^{-24}	yocto	y

Derived SI units

Quantity	Unit name	Symbol	Definition
Area (A)	square meter	m^2	$m{\cdot}m$
Volume (V)	cubic meter	m^3	$m{\cdot}m{\cdot}m$
Speed or velocity (v)	meters per second	ms^{-1}	ms^{-1}
Force (F)	newton	N	$kg{\cdot}m\,s^{-2}$
Energy (E), work (w), heat	joule	J	$N{\cdot}m(m^2{\cdot}kg\,s^{-2})$
Power	watt	W	$J\,s^{-1}(m^2{\cdot}kg\,s^{-2})$
Pressure (P)	pascal	Pa	$N\,m^{-2}(kg\,s^{-2}\,m^{-1})$
Frequency (ν)	hertz	Hz	cycles s^{-1}
Electric charge (Q)	coulomb	C	$A{\cdot}s$
Electric potential (ψ)	volt	V	$W\,A^{-1}(J\,A^{-1}\,s^{-1})$; $J\,C^{-1}$
Electric resistance (R)	ohm	Ω	$V\,A^{-1}$
Electric conductance (G)	siemens	S	$A\,V^{-1}(\Omega^{-1})$
Electric capacitance (C)	farad	F	$C\,V^{-1}$
Concentration	moles per cubic meter	$mol\,m^3$	$mol\,m^3$
Irradiance (energy)	watts per square meter	$W\,m^{-2}$	$J\,s^{-1}\,m^{-2}$
Irradiance (moles of photons)	moles per square meter second	$mol\,m^{-2}\,s^{-1}$	$mol\,m^{-2}\,s^{-1}$
Spectral irradiance (moles of photons)	moles per square meter second nanometer	$mol\,m^{-2}\,s^{-1}\,nm^{-1}$	$mol\,m^{-2}\,s^{-1}\,nm^{-1}$
Magnetic field strength	amperes per meter	$A\,m^{-1}$	$A\,m^{-1}$
Activity (of radioactive source)	becquerel	Bq	s^{-1}

SOME DISCARDED METRIC UNITS

Discarded metric units	Acceptable SI units
micron (μ)	micrometer (μm)
millimicron (mμ)	nanometer (nm)
angstrom (Å)	0.1 nanometer (nm)
bar (bar)	0.1 megapascal (MPa); 100 kilopascal (kPa)
calorie (cal)	4.1842 joule (J)
degree centigrade (°C)	degree Celsius (°C)
hectare (ha)	10 000 m^2 or 0.01 km^2
einstein (E)	mole of photons or quanta (mol)
parts per million (ppm)	mg kg^{-1}
	μmol mol^{-1} (e.g., CO_2 in air)
	(Use kg for mixed substances and mol for pure substances and gases)
	1000 $mm^3\,m^{-3}$ (volume; e.g. liquids)

parts per billion (ppb) $\mu g\ kg^{-1}$
                                   $nmol\ mol^{-1}$
                                   $mm^3\ m^{-3}$ (volume; e.g., liquids)

Important notes:
- Pluralize names of units when they are greater than 1, equal to 0, or less than −1; otherwise, keep them singular. Example: 1000 micrograms, 0 degrees Celsius, −10 degrees Celsius, −0.2 degree Celsius, 0.5 liter.
- Never make symbols for units plural (20 ml *not* 20 mls).
- Do not place periods after symbols unless they are at the end of a sentence.
- Never begin a sentence with a symbol or with a numeral.
- When a quantity is used as an adjective, insert a hyphen between the numeral and the unit to avoid confusion. Example: 500-watt light bulb, a 10-ml sample.
- Do not place a comma between three-digit groups; use a space instead. Example: 10 000 apples, not 10,000 apples.
- Compound units may be written in the denominator using a slash (/) with no spaces before or after the slash. Example: m/s. Compound units may also be written (and this is usually preferred) as negative exponents. Example: $m\ s^{-1}$.

Some information for appendix 3 was taken from

Salisbury, F.B., and Ross, C.W. (1992). *Plant Physiology 4th ed*. Wadsworth Publishing Co., Belmont, California, pp 601–605

Index